CAMBRIDGE COUNTY GEOGRAPHIES

General Editor: F. H. H. GUILLEMARD, M.A., M.D.

# ESSEX

*Cambridge County Geographies*

# ESSEX

by

GEORGE F. BOSWORTH, F.R.G.S.

With Maps, Diagrams and Illustrations

Cambridge:
at the University Press
1909

CAMBRIDGE UNIVERSITY PRESS
Cambridge, New York, Melbourne, Madrid, Cape Town,
Singapore, São Paulo, Delhi, Mexico City

Cambridge University Press
The Edinburgh Building, Cambridge CB2 8RU, UK

Published in the United States of America by Cambridge University Press, New York

www.cambridge.org
Information on this title: www.cambridge.org/9781107685543

First published 1909
First paperback edition 2013

*A catalogue record for this publication is available from the British Library*

ISBN 978-1-107-68554-3 Paperback

# CONTENTS

# ILLUSTRATIONS

## MAP

The Illustrations on pp. 7, 11, 23, 24, 43, 46, 48, 63, 67, 76 and 87 are from photographs by Messrs F. Frith and Co., Ltd., Reigate; and those on pp. 4, 12, 16, 19, 57, 72, 81, 83, 93, 98, 100, 102, 107, 109, 114, 115, 137, and 150 are from photographs by Mr A. Wire, Leytonstone.

# 1. County and Shire. The Word *Essex*. Its Origin and Meaning.

Essex is one of the largest of the six Home Counties, and its maritime position gives it considerable importance. It was one of the early English kingdoms, and for more than a thousand years it has represented in brief the history of England. In many of its towns and villages there are monuments of antiquity and treasures of art which help us to realise the past history of our country; and we shall understand much better the progress and development of England as a whole, if we first carefully study the geography and history of the county of Essex.

In this chapter we will consider the meanings of the two words, *shire* and *county*, and then endeavour to trace the origin of the county of Essex and the meaning of its name. For a thousand years and more the county, or shire, in England has been considered the chief unit of local government, in much the same way that the canton is regarded in Switzerland, the department in France, and the state in America. We now have the two words shire, and county, but before the Norman Conquest the word shire only was used.

In the earliest period of our history the word shire simply meant a division, and we find the word was thus used to denote the various portions of Cornwall, and the two kingdoms of Kent. Then, as time passed on, the word acquired a new meaning, and was applied to any portion that was shorn off or cut off from a larger division. The portion cut off was a share or shire, and hence many of our counties have retained this affix since the settlement of the English in our land.

The word county is due to the Norman invaders who identified the old English shire with their own *comitatus*, the district of a *comes* or count. And thus it comes about that we use the two words shire and county to denote the larger divisions of our land that were made long ages ago.

The counties of England differ considerably in their origins. Such counties as Essex, Kent, and Sussex had a different origin from Nottinghamshire, Leicestershire, and Northamptonshire. The former are probably survivals of former kingdoms, while the latter are undoubtedly shares, or shires, of the kingdom of Mercia. Essex, Kent, and Sussex have kept their names and boundaries from the earliest times, perhaps for more than fourteen hundred years, and it is the knowledge of such facts as these that makes the study of county geography and history so interesting. Indeed it is not too much to say that some of our counties are an epitome of our national history, and to illustrate the truth of this we could not have a better example than the county of Essex.

It is rather difficult to trace the origin of some of our counties and the meaning of their names, but we have no such difficulty in the present case. Here we have a county with a distinctly English name, which was derived from the Saxons who settled in this part of England. The Saxons were perhaps the strongest of the various invaders of our country in the fifth century, and they were distinguished by the portions they settled along the eastern and southern districts. Thus we read of the East Saxons, South Saxons, West Saxons and so on, while the other invaders, the Angles and Jutes, settled elsewhere. The word Essex is thus derived from the East Saxons, and in the *English Chronicle* we find it written Easte seaxe. Later it appears as East saxe, and in Domesday Book it is written Exsessa. Then the form Essex was used by our historians, and so it has continued to this day.

The East Saxons formed the kingdom of Essex in a district that had been settled by a Keltic tribe known as the Trinobantes, or Trinovantes, a word which means battle-spearers, or battle-stabbers. The territory of the Trinobantes was fairly compact, comprising as it did the modern county of Essex and a part of Middlesex, from beyond the Lea to the Stour on the north. Middlesex was included for a long period in the kingdom of Essex, and then for some reason, perhaps owing to the growth of London, it was separated from the East Saxon kingdom and made into a county. And this point of interest deserves notice here, for the East Saxon kingdom was also the see of the bishop of London, and so it continued till quite recent times, when owing to the growth of population Essex was annexed to the see of Rochester, and then to that of St Albans. Now Chelmsford is likely to be

the see of a bishop of Essex, thus linking the county with
the earliest period of our history.

Thus we may say that the modern county of Essex
grew out of the East Saxon kingdom, which was formed

Bow Bridge and Wharf

from the territory that had been occupied by the Trinobantes, a Keltic tribe living in Britain when the Romans
first landed in our country.

## 2. General Characteristics. Its Position and Natural Conditions.

Essex is a large maritime county on the east of England, and its long coast-line both along the North Sea and the estuary of the Thames is of great advantage. From at least one of its sea-ports there is constant intercourse with the Continent, and several of the smaller ports are fishing stations or yachting centres.

As an agricultural county Essex is of considerable importance, and although in recent years it has passed through a period of depression, there are signs that it will ere long take its former rank as one of the leading agricultural counties. Notwithstanding the long succession of lean years, about two-thirds of the county is now under cultivation, and on every hand it is evident that Essex farmers are now working in a more scientific way, and are turning their attention to dairy farming and fruit farming on new principles.

Essex has little claim to rank as an industrial county. It is true that many minor industries are carried on in some of the towns, and that the extensive railway works at Stratford employ several thousand men, but there is no staple industry such as flourishes in Yorkshire or Lancashire.

Next to its importance as an agricultural county, Essex ranks as one of the metropolitan counties, and thus shares with Surrey and Kent the many advantages arising from the proximity of London. That portion near London, which lies in the hundred of Beacontree, has considerably more than half the population of the whole county. West Ham, East Ham, Leyton, Walthamstow, Barking, Ilford, Wanstead, and Woodford—all these places lie close together, and have been well called London-over-the-Border. Here it is that a large number of the workers of London live. The city man, the clerk, the artisan, and the labourer leave London every evening and travel to their homes in Kent, Essex, and Surrey, and of these many thousands Essex claims a large proportion.

This London-over-the-Border in Essex is unlike any other part of the county. Forty or fifty years ago it was a district of green fields and pretty gardens, of farms and orchards. Each parish had its large houses owned by rich merchants and bankers who drove to and from London in their coaches and carriages; while the cottagers dwelt in pleasant homes with many of the advantages of a country life. Now there are countless miles of streets with commonplace houses of identical pattern, while most of the large houses have disappeared and the rich people have retreated further into the county. This suburban portion of Essex with its immense population forms but a small part of the whole county, which fortunately retains its rural character.

Essex is a county that needs to be known to be appreciated, and although it has not so many charming spots as Surrey, there is much that pleases the traveller in search of pastoral and beautiful scenery. Many writers have described it in glowing terms, and from the time of John Norden, we have ample evidence that it has com-

White Notley Village

mended itself to the most diverse and exacting minds. Norden, who lived in Elizabeth's reign, says "This shire is moste fatt, frutefull, and full of profitable things, exceding (as far as I can finde) anie other shire, for the generall comodeties, and the plentie...this shire seemeth to me to deserve the title of the Englishe Goshen, the fattest of the Lande : comparable to Palestina that flowed with milke and hunnye." Perhaps Norden's description is rather flattering, but it is interesting from the fact that it was written by a very competent observer more than three hundred years ago.

Arthur Young, who knew well the England of the eighteenth century, declares that some of the Essex scenery was equal to anything "even in the west of England, that region of landscape." A modern writer on the England of to-day has recorded his impressions of Essex in "A Tour in a Phaeton through the Eastern Counties." Mr Hissey, in that book, is astonished that "spots of so much beauty should be so near town and so little known." He sums up the chief points that attracted his attention while travelling through Essex as follows :—"Ruined abbeys and ancient churches fraught with interest for the ecclesiologist and antiquary ; romantic homes of the old days —many of these moated still—all abounding in past memories and historic associations; old-time coaching hostelries wherein our forefathers made merry ; old fashioned, oddly built country towns; picturesque hamlets ; pleasant pastoral scenes varied by wild wind-swept heaths and gorse-sprinkled commons." Here, it must be admitted, we have enough to commend Essex to our attention, and

Lees Priory

we shall find in this book a confirmation of these statements that Essex is indeed a county with many attractions
and a charm of its own.

### 3. Size. Shape. Boundaries. De-tached Portion. Transfer of Parishes.

In the previous chapters we have learnt how Essex
came to be a county and of its characteristics. It will now
be well to consider its size, shape, and boundaries. At
the outset it may be noted that the ancient county of
Essex is somewhat larger than the administrative county,
as there are portions which are included in Cambridge and
Suffolk, and there is a small area which belongs to Kent
but is now included in the county of London.

Essex may almost be called a peninsula, for it has water
on four sides. In shape it is an irregular quadrilateral
having a perimeter of about 225 miles, if we measure the
windings of the rivers and the sea-coast. From Stratford
in the south-west to Harwich in the north-east the diagonal
measures 63 miles ; from Bartlow in the north to Tilbury
in the south the length is over 50 miles; and from Roydon
in the west to Walton-on-the-Naze in the east the breadth
is about 60 miles.

The area of the ancient county of Essex is 986,975
acres, and of the administrative county 979,532 acres. If
we take the former area as equal to 1542 square miles, we
find that Essex is about one-thirty-third of the whole of

England.   There are nine ancient counties larger than
Essex, which ranks just below Kent, but above Suffolk in
area.

When we consider the boundaries of our county, we
are not confronted with the same difficulties that often
perplex us in trying to determine those of other counties.

Walton-on-Naze

If we look at a good map, the first thing that strikes us
is that Essex has natural boundaries on all sides, except
a small portion on the north-west.   The northern
boundary is the river Stour from its source at Haverhill
to the end of its wide estuary at Harwich.   The North
Sea washes Essex on the east, and the estuary of the river
Thames from Stratford to Shoeburyness is the boundary

on the south. The river Lea enters Essex from Hertfordshire at Roydon, and from that town southwards to its
junction with the river Thames, it forms the western
boundary. From Haverhill to Bishop Stortford the
boundary is artificial. The counties of Suffolk and Cambridge are on the north; Hertfordshire and Middlesex
are on the west; and Kent is on the south of Essex.

Flatford Bridge

The size of the present county of Essex was doubtless
about that of the ancient kingdom of the East Saxons,
who were separated from the kingdom of East Anglia by
the river Stour, and from the kingdom of Kent by the

river Thames.  The river Lea formed the boundary between the later East Saxon kingdom on the one side and Middlesex and Mercia on the other.  Perhaps few counties in England have altered so little as Essex, both as regards size and boundaries, since the earliest period of our history.

It used to be quite a common thing for a portion of one county to be included in another county.  There are several counties where this is still the case, but the tendency of recent years has been to remove this anomaly.  There is, however, one portion in the south of Essex which belongs to the ancient county of Kent.  This district is known as North Woolwich and for administrative purposes it belongs to London.  There is no really satisfactory reason to be given why a portion of Kent should be in Essex.  Some historians think that this isolated portion dates from the time when Essex and Kent were one kingdom : and that when they were separated it was arranged that Kent should retain this territory in Essex.  This is plausible, but another reason has found more favour.  It appears that Count Haimo, Sheriff of Kent in William the Conqueror's reign, had land on both sides of the Thames at Woolwich, and in this way his possessions on the north bank in Essex became included in the county of Kent.

A few changes have been made in Essex during the last twenty years by the transfer of some parishes in the north and north-west to the adjoining counties.  In 1888, Ballingdon-cum-Brandon was given to Suffolk ; and, in 1894, Great and Little Chishall and Heydon were transferred to Cambridge, and Kedington to West Suffolk.

## 4.  Surface and General Features.

The surface and structure of Essex are more easily understood if we consider it first of all as part of a larger area.  Its physical structure is the same as that of the East Anglian counties, and the three counties of Norfolk, Suffolk, and Essex belong to the eastern plain of England. A glance at the map of England will show that these counties form a quadrant, and are severed almost as completely from the rest of England as the three counties of the Wealden plain—Kent, Surrey, and Sussex.

It is interesting to note that Norfolk, Suffolk, and Essex have been shaped round the estuaries upon which the original settlements were formed, and the chief towns in each county are on the rivers near the sea.  All the chief rivers of the three eastern counties rise in the highland which is mostly situated in the west, and flow in a generally eastwardly course to the North Sea.

There is one point of contrast to which we must refer while we are dealing with the structure of the East Anglian counties.  The chalk which forms so great a feature in the other eastern counties only passes over the north-western portion of Essex on its way to Hertfordshire, while the strip of alluvial soil which edges the eastern counties becomes in Essex a broad belt of clay extending over the greater portion of its surface.

In dealing with the physical features of Essex we shall not be far wrong if we say that about seven-tenths of its surface is low, in many parts on a level with the sea, and in some parts even below the sea. Let us then consider the surface of the county under five divisions—the chalk uplands, the Essex heights, the forests, the lowlands, and the marshes.

The chalk uplands occupy a fairly large area in the north-west of the county ; and the county westward of a line drawn from Bishop Stortford on the Lea to Birdbrook near the Stour would comprise this hilly region. In parts the appearance of this district resembles that of the South Downs, though it is not so undulating. The well-rounded hills are separated by the valleys, more or less deep, and in the north-west corner some ten square miles rise to more than 400 feet. The highest point in Essex is a short distance north of Langley church, where the elevation is 485 feet.

The Essex Heights extend from Colchester to Purfleet on the Thames, and include three hilly districts, Tiptree Heath near Witham, Danbury Hill near Chelmsford, and Laindon Hill near Billericay. Tiptree Heath does not reach 300 feet, and is the centre of a district that was once an extensive wild woodland, broken here and there with heather wastes. The heath is now mostly enclosed and cultivated, and has lost nearly all of its ancient wildness and rough beauty. The highest point of Danbury is on the road above Lingwood Common, 353 feet above the sea. Although we will not say the scenery is grand in the Danbury district, we can safely call it really charming. The view from Danbury Hill is a bold and sweeping panorama, full of wood and water, parks and villages. On the one hand lies the sea, while on the other there is a spreading landscape of typical English scenery. Laindon Hill, rising to a height of 386 feet, stands by itself far from any other hills. The prospect from it is one worth going far to see, and Arthur Young considered it the finest view in England. This, of course, is exaggeration, but from the summit there is a glorious view of

Epping Forest.  Typical Forest Scenery

(*Pollarded Old Hornbeams in Honey Lane Vale, High Beech*)

waving woods, green meadows, and well-tilled fields, while in the far distance we get a glimpse of the Thames and the hills of Kent.

The forests of Essex now consist chiefly of Epping Forest and Hainault Forest, which are the remains of the Forest of Essex, or, as it was called in later times, Waltham

Forest. This district of over 6000 acres is perhaps the most picturesque and important feature of the county. It is well-wooded and undulating, reaching its greatest height at Epping, which is 381 feet above the sea. High Beech is the chief summit, commanding an extensive view across the valley of the Lea into Middlesex and Hertfordshire.

The lowlands of Essex comprise the larger portion of the county, and occupy, with the exceptions just mentioned, the whole of the area from the chalk uplands of the north-west to the marshes of the coast. Of course the characteristics of the Essex lowlands vary, yet broadly speaking it is a great agricultural region on the London Clay.

The Essex marshes are along the sea-coast and the Thames estuary. They are in many cases extensive tracts of land reclaimed from the sea and protected by sea-walls, constructed of mud and stones. These marshes are chiefly under grass, and are intersected with numerous wide ditches known as "fleets." At one time the Essex marshes had a very evil reputation, but by draining and other methods of management they are now far more healthy than they were even during the last century.

What are known as the Essex "Saltings" are some alluvial tracts which comprise not only the old embanked area below high-water mark, but also salt marshes, which rise 10 feet and more above ordnance datum. The spring tides cover these "saltings," and by leaving thin films of sediment tend gradually to raise their level, until in the end the sea may be excluded. As the saltings widen seaward, fresh strips have been enclosed. The great trouble to residents in the Essex marshes has hitherto been the want of fresh water, but now deep wells sunk through the London Clay have gone far to remedy this drawback.

## 5. Watershed. Rivers—Thames, Stour, Lea, Roding, Bourne Brook, and Ingerburn.

Essex is a well-watered county, and as the slope of the county is to the south-east, it will be seen that most of the longer streams flow in that direction. The chief watershed is in the chalk downs in the north-west, although there is a ridge of high ground in the south, in which the Crouch and some other streams rise. There are several features that are common to the chief Essex rivers. A glance at the map will show that the principal streams find their way through a level and often marshy country to the North Sea. Each of them has an estuary, in which or near which there are islands, most of them being separated by a very narrow channel from the mainland.

We may classify the rivers of Essex as boundary rivers, the Thames, the Lea, and the Stour; and as internal rivers, the Colne, the Blackwater, the Chelmer, the Crouch, the Roding, the Bourne Brook, the Ingerburn, and the Cam. Of course, the Thames cannot be called an Essex river, but as it bounds the southern shore of the county it will be specially noticed in the chapter on the coast of Essex.

Here we need only remark that the Thames begins to bound Essex from the point where the Lea joins it ; and the estuary of the Thames extends to the Nore lightship. From London to the Nore " the Thames is the world's exchange," and although this part of the river is not beautiful as regards its scenery, its commercial importance is unrivalled.

Harwich. Boat leaving Pier

The Stour, forming the boundary between Essex and Suffolk, rises not far from Haverhill on the border of Cambridgeshire. The river soon expands and becomes a lake, or *mere*, giving its name to the parish of Sturmer or

Stourmere. After receiving two small tributaries, it passes Sudbury and widens till it meets the tide at Manningtree. From that town to the sea it forms an estuary ten miles long, and as much as one mile wide. Harwich stands at the entrance to the estuary on the south, or Essex side, and it is at this port that the Orwell also reaches the sea. From Landguard Fort on the north to Harwich is a fine expanse of water, but owing to the set of the tidal currents and other causes, the banks have a tendency to increase, thus rendering the harbour of Harwich less easy of approach. The country through which the Stour flows is of a quiet, pastoral character; much of it is typical English landscape, which Constable and Gainsborough loved to paint.

The Lea, or Lee as it is sometimes written, is the boundary river on the west, separating Essex from Middlesex and Hertfordshire. The name of the river is of English origin ; and although the Lea is not now of much importance as a commercial highway, it has played an important part in the history of our land from the time of Alfred downwards. Michael Drayton sang its charms, and Izaak Walton walked by its banks and wrote lovingly of its value from the angler's point of view. It may not afford many very striking features of landscape scenery, yet it presents several of considerable charm. When Izaak Walton "sat on a primrose bank and looked down the meadows," he thought "they were too pleasant to look on but only on holidays," and then adds :

> " I in these flowery meads would be,
> These crystal streams should solace me."

The Lea is not entirely an Essex river, for it rises at
Houghton Regis, not far from Dunstable, in Bedfordshire.
Entering Hertfordshire near Harpenden, it proceeds in a
south-easterly direction through Hatfield Park, and then
inclining to the north-east it passes Hertford and Ware.
From the latter town it bends again to the south, and is
joined by the Stort near Roydon, from which place it
forms the boundary between Essex and Hertfordshire, till
it leaves the latter county near Waltham Abbey.   It then
separates Essex from Middlesex, continuing to flow nearly
south, till it enters the Thames at Blackwall, opposite
Greenwich marshes.   By means of cuttings it has been
rendered navigable for barges as far as Hertford ; but it
was once a much larger river than it now is, for more
than 1000 years ago Alfred diverted its waters to lay the
Danish fleet aground.   There is every reason to suppose
that from Chingford onward to its junction with the
Thames, the Lea had a very considerable estuary, and
Danish boats have been dug up in recent years in the
neighbouring marsh-lands.

The lower portion of the Lea from Chingford is now
marked by a series of 12 great reservoirs having a water
area of 479 acres, a total storage capacity of 2,400 million
gallons, and a shore line of over 15 miles.   These are
the reservoirs of the Metropolitan Water Board, and
furnish the supply of water to a population of nearly
two million people.   Every day upwards of 40 million
gallons are conveyed to this enormous population, through
1036 miles of pipes.   Ten of these reservoirs can be filled
by gravitation ; and two can be so filled up to half of

their capacity, and the remainder by pumping. Besides these open reservoirs, there are several covered reservoirs, numerous pumping stations, and three large filter beds. Owing to the large and increasing population of London-over-the-Border and East London, it is contemplated to construct other reservoirs so as to meet the great demands of this district.

After the Lea, the Roding is the principal tributary of the Thames in Essex. Its name used to be written Roden, and Roothing, but neither of these forms is now used. The Roding rises in Easton Park, near Dunmow, and passes through an agricultural district known as "the Roothings," which comprises eight parishes bearing this name. The river is joined by numerous little streams on its southward journey, which lies through fertile tracts of meadow land, often flanked by belts of woodland and rich cornfields. The length of the Roding from its source to its junction with the Thames at Barking Creek is about 34 miles. During a wet season floods often occur in its lower course, and if the water from the forest above is met by the tidal water from below, the banks or "walls," as they are called, give way, the meadows are flooded, and much damage is the result. The tide ascends beyond Ilford, and the Roding has been embanked as far as that town.

The two little feeders of the Thames—the Bourne Brook and the Ingerburn—need not detain us long. The Bourne Brook has its source in the high ground to the south of Navestock, and flows past Romford to the Thames. The Ingerburn rises not far from the Bourne in the neighbourhood of Brentwood. Thence it flows southward by Upminster and Rainham, and after a short course of 12 miles joins the Thames.

## 6. Rivers—Colne, Blackwater, Chelmer, Crouch and Cam.

In the last chapter we read about the border rivers of Essex, and about the tributaries of the Thames in this county. We have now to consider the remaining rivers whose courses are entirely within the county.

The River Colne at Colchester

Let us begin with the river Colne, which is the most important river in the north. It takes its name from *Colonia*, the Roman Colchester, and in this respect differs

from most of our rivers, which often give their names to
the chief towns.   The Colne rises near Birdbrook in the
north-west of Essex, not far from the source of the Stour.
It flows south-east through a picturesque district, passing
Yeldham and Hedingham on its way to Halstead and
Colchester.   At Wivenhoe it receives a little stream, the
Roman River, and then forms an estuary, on which stands

The Bath Wall, Maldon

Brightlingsea, a yachting centre.   From its source to
Colne Point the length of the Colne is about forty miles,
and it is navigable for small craft as far as Colchester.

Proceeding southwards we come to the Blackwater,
which is called the Pant in the earlier part of its course.
Its source is at Wimbish, not far from Saffron Walden, in

the hilly district of north-west Essex. Its general direction is south-east through a purely agricultural country. The chief towns it passes are Coggeshall, Kelvedon, and Maldon. At the latter town it enters the sea by a large estuary to which it gives its name, and which also receives the water of the Chelmer. Four miles above its junction with the Chelmer, the Blackwater receives on its right bank Pods Brook, a stream which passes Braintree and Witham. The estuary of the Blackwater is a tidal channel ten miles in length; but the navigation of the river does not extend beyond the junction of the Chelmer.

The Chelmer, which gives its name to the county town, rises a short distance south of the Blackwater and flows south-east by Thaxted, Dunmow, and Felstead to Chelmsford. At the county town it receives the river Cann, which rises near High Roding. The Chelmer turns almost due east at Chelmsford and so continues to its junction with the Blackwater. Towards the end of the eighteenth century, the river was canalised from Chelmsford, and by means of numerous locks was made navigable as far as Maldon, a distance of fifteen miles. Both the Blackwater and the Chelmer have a length of nearly forty miles.

The Crouch is the most southerly of the Essex rivers. It rises in two little streams near Little Burstead and Laindon, and has an eastward course of nearly twenty miles on its way to the North Sea. The Crouch drains a district which is somewhat different from any other part of Essex, but which has a quiet beauty of its own. It is often known as the "Battle" River, from the fact that Canewdon, Hockley, and Ashingdon claim to have been the scenes of battles in the reign of Canute. The estuary of the Crouch extends for a distance of thirteen miles, and opposite Burnham, which is a capital yachting centre, there is quite an archipelago of islands. This broad tidal river flowing between low hills on either side and carrying hundreds of pleasure yachts forms a most picturesque scene, quite unique of its kind.

The last river we shall notice is the river Cam, which has the smallest river basin in the county. It is composed of two branches, one rising in Bedfordshire, while the other, bearing the classic name of the Granta, has its source in Quendon among the chalk hills in the north-west of Essex. It then flows through the beautiful grounds of Audley End, and leaves the county at Chesterford, on its way to Cambridge.

## 7. Geology and Soil.

By Geology we mean the study of the rocks, and we must at the outset explain that the term *rock* is used by the geologist without any reference to the hardness or compactness of the material to which the name is applied; thus he speaks of loose sand as a rock equally with a hard substance like granite.

Rocks are of two kinds, (1) those laid down mostly under water, (2) those due to the action of fire.

The first kind may be compared to sheets of paper one over the other. These sheets are called *beds*, and such beds are usually formed of sand (often containing pebbles), mud or clay, and limestone or mixtures of these materials. They are laid down as flat or nearly flat sheets, but may afterwards be tilted as the result of movement of the earth's crust, just as you may tilt sheets of paper, folding them into arches and troughs, by pressing them at either end. Again, we may find the tops of the folds so produced washed away as the result of the wearing action of rivers, glaciers and sea-waves upon them, as you might cut off the tops of the folds of the paper with a pair of shears. This has happened with the ancient beds forming parts of the earth's crust, and we therefore often find them tilted, with the upper parts removed.

The other kinds of rocks are known as igneous rocks, which have been molten under the action of fire and become solid on cooling. When in the molten state they have been poured out at the surface as the lava of volcanoes, or have been forced into other rocks and cooled in the cracks and other places of weakness. Much material is also thrown out of volcanoes as volcanic ash and dust, and is piled up on the sides of the volcano. Such ashy material may be arranged in beds, so that it partakes to some extent of the qualities of the two great rock groups.

The production of beds is of great importance to geologists, for by means of these beds we can classify the rocks according to age. If we take two sheets of paper, and lay one on the top of the other on a table, the upper one has been laid down after the other. Similarly with two beds, the upper is also the newer, and the newer will remain on the top after earth-movements, save in very exceptional cases which need not be regarded by us here, and for general purposes we may regard any bed or set of beds resting on any other in our own country as being the newer bed or set.

The movements which affect beds may occur at different times. One set of beds may be laid down flat, then thrown into folds by movement, the tops of the beds worn off, and another set of beds laid down upon the worn surface of the older beds, the edges of which will abut against the oldest of the new set of flatly deposited beds, which latter may in turn undergo disturbance and renewal of their upper portions.

Again, after the formation of the beds many changes may occur in them. They may become hardened, pebble-beds being changed into conglomerates, sands into sand-stones, muds and clays into mudstones and shales, soft deposits of lime into limestone, and loose volcanic ashes into exceedingly hard rocks. They may also become cracked, and the cracks are often very regular, running in two directions at right angles one to the other. Such cracks are known as *joints*, and the joints are very important in affecting the physical geography of a district. Then, as the result of great pressure applied sideways, the rocks may be so changed that they can be split into thin slabs, which usually, though not necessarily, split along planes standing at high angles to the horizontal. Rocks affected in this way are known as *slates*.

If we could flatten out all the beds of England, and arrange them one over the other and bore a shaft through

|  | Names of Systems | | Characters of Rocks |
|---|---|---|---|
| TERTIARY | Recent & Pleistocene | | sands, superficial deposits |
| | Pliocene | | |
| | Eocene | | clays and sands chiefly |
| SECONDARY | Cretaceous | | chalk at top<br>sandstones, mud and clays below |
| | Jurassic | | shales, sandstones and<br>oolitic limestones |
| | Triassic | | red sandstones and marls, gypsum and salt |
| PRIMARY | Permian | | red sandstones & magnesian limestone |
| | Carboniferous | | sandstones, shales and coals at top<br>sandstones in middle<br>limestone and shales below |
| | Devonian | | red sandstones,<br>shales, slates and limestones |
| | Silurian | | sandstones and shales<br>thin limestones |
| | Ordovician | | shales, slates,<br>sandstones and<br>thin limestones |
| | Cambrian | | slates and<br>sandstones |
| | Pre-Cambrian | | sandstones,<br>slates and<br>volcanic rocks |

them, we should see them on the sides of the shaft, the newest appearing at the top and the oldest at the bottom, as shown in the figure. Such a shaft would have a depth of between 10,000 and 20,000 feet. The strata beds are divided into three great groups called Primary or Palaeozoic, Secondary or Mesozoic, and Tertiary or Cainozoic, and below the Primary rocks are the oldest rocks of Britain, which form as it were the foundation stones on which the other rocks rest. These may be spoken of as the Pre-Cambrian rocks. The three great groups are divided into minor divisions known as systems. The names of these systems are arranged in order in the figure with a very rough indication of their relative importance, though the divisions above the Eocene are made too thick, as otherwise they would hardly show in the figure. On the right hand side, the general characters of the rocks of each system are stated.

With these preliminary remarks we may now proceed to a brief account of the geology of the county.

Essex is part of a tract known as the London Basin, which has for its framework the chalk formation. The south rim of this basin comes to the surface in Essex at Purfleet and Grays, and the north rim appears in the chalk uplands of north-west Essex at Heydon, Saffron Walden, and elsewhere. The hollow of the basin is known as the London Clay, which covers about four-fifths of the county. The succeeding deposits in Essex are chiefly small outlying tracts of crag, and some accumulations of glacial drift. These lie over the uplands as gravel and boulder clay, and in river valleys in the form of gravel, brick-earth, and alluvium. From this brief outline it will be seen that the geology of Essex is not so varied as in some counties, and by the aid of the geological map it will be quite easy to understand the following details.

The most ancient rocks known to geologists are the Primary or Palaeozoic, but none of them come to the surface in Essex. In 1858, as a result of boring to a depth of 1029 feet below the surface at Harwich, the coal measures were not reached, but there were evidences of some slaty rock. With this exception, the oldest formation in Essex is the gault, which consists of stiff blue and grey clay and marl, having a thickness of 72 feet at Loughton. Although the gault does not appear at the surface in Essex, it is probably everywhere present beneath the chalk.

The chalk is seen at the surface over a small portion of Essex, and where it does occur its presence is indicated by pits and lime-kilns. What is known as the middle chalk is exposed in the neighbourhood of Heydon, Great Chesterford, and Hadstock in the north-west. It is a hard rock chalk, about ten feet thick, well-bedded in layers, and with very few flints. The upper chalk appears above ground at Grays and Purfleet in the south, and in parts of the north-west. It is generally soft and has layers of flint. At Hangman's Wood near Grays there are some remarkable excavations in the chalk known as Deneholes. Shafts are carried through fifty or sixty feet of gravel, and then continued about twenty feet into chalk. Who made these Deneholes, with their remarkable and extensive chambers, and what purpose they were intended to serve, will be considered in another chapter. The chalk is important as a water-bearing formation, and from it supplies are obtained for many of the Essex wells.

Overlying the chalk there is generally found a mass of pale and greenish-grey sand, and sandy clay. This formation is known as the Thanet Beds, which are exposed between Purfleet and Aveley, at Stifford, Chadwell, and elsewhere. The Thanet Beds are succeeded by the Woolwich and Reading Beds, which consist of mottled clay, clay and sand, greenish-grey sand, and flint pebbles. These beds are from twenty-five feet to sixty feet in thickness and are seen at the surface at Orsett and Stanford-le-Hope in the south, and at Thaxted and Castle Hedingham in the north.

After the Blackheath Beds, which have been recognised at Barkingside and Shoeburyness, we come to the London Clay. This formation, upwards of 400 feet thick, may be seen exposed in the cliffs at Southend, Clacton, Frinton, and Walton-on-the-Naze. It occurs at the surface along the valley of the Crouch, and over Epping and Hainault Forests, in the south of Essex, and along the valleys of the Colne, Blackwater, and Chelmer. It is a stiff brown clay which soaks up a good deal of water in wet weather, and shrinks and cracks in very dry weather. The London Clay is extensively dug for brick and tile making in South Essex.

The Bagshot Beds are found over the London Clay and consist of fine, light-coloured sands with layers of pipe clay, and, in places, pebble beds. This formation occurs in outliers at Epping, High Beech, Brentwood,

Weeley, and Laindon Hill. In the neighbourhood of the Bagshot Beds the land is less cultivated, and commons, village greens, and woodlands diversify the scenery.

Following the Bagshot Beds comes the Red Crag, one of the most attractive of geological formations, for it is found in pleasant places, and from it fossils can readily be obtained. In Essex the Red Crag is not widely distributed, and its most famous locality is at Walton-on-Naze.

Valley gravel and brick-earth occur chiefly in the Thames valley from Stratford and Leyton to Barking and Romford. At Ilford and Grays this brick-earth has been extensively worked, and at Uphall numerous mammalian remains have been found. These include bones of the hippopotamus, rhinoceros, Irish elk, and red deer.

We now come in order of geological time to the recent deposits in Essex, which may be named alluvium. These deposits are a series of muds or silts, clays, and peat with sand or gravel, having a thickness of as much as fifty feet at Thames Haven in the south. The alluvium in Essex is really the tract of marshland borderi g the river estuaries and the inlets between Walton and Harwich, and is in some places six or seven feet below the level of high water at spring tides. At Walthamstow there is a breadth of this alluvium forming the extensive marshes by the side of the Lea. In 1868–9 some large reservoirs were being constructed for the East London Water Company in this district. Among the many interesting remains that were discovered there were bones of a few birds and fishes, skulls of man, implements of stone, bone, ivory, and iron, and portions of the skeletons of the dog, fox, horse, deer, reindeer, elk, and beaver.

Although Essex is generally flat, and especially along the river and sea-coasts, which are bounded by the marshes, yet inland there is an abundant variety of soil, and many parts of the county are pleasantly diversified with arable land and pasture, with woodlands, commons, and well-timbered hedgerows.

## 8. Natural History.

Various facts, which can only be shortly mentioned here, go to show that the British Isles have not existed as such, and separated from the Continent, for any great length of geological time. Around our coasts, for instance, are in several places remains of forests now sunk beneath the sea, and only to be seen at extreme low water. Between England and the Continent the sea is very shallow, but a little west of Ireland we soon come to very deep soundings. Great Britain and Ireland were thus originally part of the Continent, and are examples of what geologists call continental islands.

But we also have no less certain proof that at some anterior period they were almost entirely submerged. The fauna and flora thus being destroyed, the land would have to be restocked with animals and plants from the Continent when union again took place, the influx of course coming from the east and south. As however it was not long before separation occurred, not all the continental species could establish themselves. We should thus expect to find that the parts in the neighbourhood of the Continent were richer in species, and those furthest off poorest, and this proves to be the case both in plants

John Ray, F.R.S.

and animals. While Britain has fewer species than France or Belgium, Ireland has still less than Britain.

3—2

The flora of Essex is smaller than that of Kent, owing no doubt to the greater uniformity of its soil. The long line of sea-coast with its extensive salt marshes might have been expected to produce more rarities than are found in it; for although there are a few maritime plants of a local character, there is generally a great sameness in the vegetation of these marshes.

Epping Forest, a fine woodland tract of nearly 6000 acres, is, for its extent, rich in its flora. This is owing to its variety of soil and aspect, and the contrast of extremely dry positions with damp hollows. Among the flowers which are either rare or remarkable, may be mentioned the sundew, the grass of Parnassus, the bog bean, and the lily of the valley.

One of the most interesting plants of Essex is the true oxlip, a very beautiful primula which is abundant in that part of the county that borders on Suffolk, Cambridge, and Hertfordshire. Essex, however, has not many rarities in its flora, but three plants may be noted, viz.—the sickle-leaved hare's ear, the Fyfield pea, and the small-fruited goose grass.

There is much that might be said of the trees of Essex, especially those growing in Epping Forest. The hornbeam, oak, hawthorn, birch, beech, are among the better known, but the hornbeam is perhaps the most distinctive of them. Epping Forest is a wood of hornbeams, which are not common in Great Britain save in Essex and Hertfordshire. The hornbeam grows in all soils, and it is said that it was originally planted for the sake of the deer, who browse on its leaves.

The wild animals of Essex are similar to those that are found in most English counties. The fox, stoat, and weasel are common throughout the county, but the badger, marten, and polecat are now rare. The squirrel is not well represented, although it is increasing in numbers. The otter is to be found in several of the Essex rivers.

The fallow deer is the most conspicuous and distinctive of the wild animals living in Epping Forest, where it has wandered for many centuries. It has remained unchanged in type and is almost a distinct breed. The deer are small in size and of a uniform dark brown, never spotted with white or fallow coloured. They number about 130, and are decreasing. The red deer is the largest and handsomest of the deer tribe of Great Britain, and was common in the Forest till the early part of the nineteenth century. It has been re-introduced in recent years, but is so mischievous that it is not looked upon with favour. The roe-deer, the smallest and most beautiful of our deer, had become extinct, but was re-introduced into the Forest by Mr E. N. Buxton in 1883, and there are now some forty of them.

Essex is rich both in the species of birds, and also in the abundance of individuals. The county is well placed, being near the Continent and on the highway followed by the migrants across the North Sea. Its coast-line is specially suited to attract all kinds of shore-loving birds, and a large extent of the surface, although it lacks mountainous tracts and large moors, has the advantage of Epping Forest and Hainault Forest, which are fine preserves for all kinds of warblers.

Among the sea-birds the black-headed gull, called in Essex the peewit, is very numerous. It is a fact worth noting that no less than three islands round the coast are named Pewit, probably because they are the breeding places of this bird. In the summer, the shores are almost devoid of bird life, for there are no rocks and cliffs; but in the autumn and winter there are swarms of gulls, divers, grebes, petrels, guillemots, ducks, and geese.

Throughout Essex the nightingale is found in every woodland. The common wagtail and kingfisher frequent the streams, and rooks congregate on the borders of the Forest. Heronries may still be seen at Birch, Goldhanger, and in Wanstead Park. The birds of Epping Forest formerly suffered much at the hands of bird-catchers, but fortunately they are now prohibited by the by-laws.

## 9. The Coast—from Bow Creek to Southend.

The coast-line of Essex is of considerable interest, for it is irregular and indented; it has numerous river-mouths and creeks; off the Crouch are many large islands; and in the shallow waters that border the east coast there are extensive sand-banks. In this and following chapters we shall deal with these subjects, and also make reference to the loss and gain along the coast, the protection of the coast, and finally the methods of lighting it.

Perhaps we shall do well in this chapter if we make a visit to the chief places round the coast commencing at

the junction of the river Lea with the Thames near
Blackwall. Essex begins at Bow Creek, and stretching
along the shore for two or three miles are the Victoria
and Albert Docks, which give accommodation for the
largest vessels. They have fine modern hydraulic cranes
and hoists, and the warehouses store immense quantities
of foreign produce. Behind the docks are the levels of
Plaistow and East Ham, which extend to the populous
borough of West Ham and the town of Barking.

Barking is one of the oldest towns of Essex, and until
recent years its fishing-smacks used to convey their supplies
of fish to Billingsgate Market. The modern town of
Barking is not at all pleasant, and the Roding, so charming
in its upper course, here becomes sluggish and muddy.
A glance at the map will show that London is much
indebted to Barking, for the great northern outfall of the
London main-drainage system is at the mouth of Barking
creek, or Creekmouth as it is called. This great sewer,
after crossing the East Ham Level, enters an enormous
reservoir at Creekmouth, whence it is discharged into the
Thames a few hours after high tide.

The course of the Thames to the sea is marked on
the Essex side by low marshes protected by walls and
embankments. Dagenham, which we soon reach, is
famous in our history for its contest with overwhelming
high waters, and we shall refer to what is known as
Dagenham Breach in another chapter. As we pass along,
we find that each bend of the Thames has its own name.
From Woolwich Reach we pass to Barking Reach, thence
to Halfway Reach, and on to Erith Reach.

The monotonous level of the marshes is broken by the chalk cliffs that appear at Purfleet. Numbers of barges are employed in carrying away cargoes of chalk and lime. Purfleet is famous for its great powder-magazine, which stores about 60,000 barrels of gunpowder. Beyond Purfleet we come to the Thurrocks—West Thurrock, Grays Thurrock, and Little Thurrock. Grays is the most important of the three, and has excellent anchorage for ships. The training ship "Exmouth" is stationed here, and there is frequent communication with Gravesend on the Kentish shore.

Leaving Little Thurrock we soon arrive at Tilbury, which is exactly opposite Gravesend. We generally associate Tilbury with the visit of Queen Elizabeth in 1588, but it had been famous long before that date. At least two hundred years before 1588, there had been some fortifications for the defence of the river. Henry VIII built a block-house, and erected a beacon, so that it is quite obvious why our English troops were centred at Tilbury when the Spanish Armada threatened us with invasion. Since the defeat of the Armada, Tilbury Fort has been enlarged and strengthened, and at the present time it is surrounded by a deep and wide ditch. While its guns command the river, its garrison, in case of necessity, could flood the whole district.

The modern Tilbury is famous for its fine and spacious docks, which are connected by the Midland Railway with London. Besides being provided with deep-water and dry docks, there are baggage sheds and extensive warehouses. The Tilbury Docks are the place of departure of the Peninsular and Oriental steamers. At East Tilbury the river takes a sharp bend to the north, and broadens considerably as it approaches the sea. Thames Haven is a place devoted to the unloading of such kinds of petroleum

The Crow Stone, Leigh

as are of too dangerous a character to bring further up the Thames. There are powder, acetylene, and gasogene works close by, so that the district is not inviting.

Beyond Thames Haven we come to a creek running

inland to Pitsea and Benfleet, and reminding us of the Danish pirates who once landed here and ravaged the country for miles around. These creeks are a common feature on the Essex coast, and were exactly suited to the shallow boats of the Danes. Passing this creek we see the low island of Canvey, which is described in Chapter 11, and then the quaint little fishing-town of Leigh comes in sight. Leigh is the last place on the Essex coast that may be said to be in the Thames, for a boundary stone, or the "Crow stone," as it is called, marks the boundary of the Thames Conservancy Board's authority.

Leigh is rapidly changing in character, and the old-world fishing-village under the cliffs is being superseded by the modern Leigh on the cliffs. The large and ancient parish church, with its massive square tower, has memories of Leigh's worthy seamen within its walls, and such notable members of Trinity H use as Salmon and Haddock will not soon be forgotten. Leigh is fast becoming a popular holiday resort, and is now connected by marine walks with Westcliffe-on-Sea, the pleasant westward extension of Southend.

The cliffs from Leigh to Southend are of some elevation, and being more or less wooded are not the least picturesque part of the Essex coast. Southend has made wonderful strides of late years, and now offers numerous attractions to the thousands of summer visitors. The view across the estuary of the Thames is both extensive and animated, and the familiar pier stretching for more than a mile and a quarter into the river is one of the sources of municipal wealth. Southend has gained a reputation through the

enterprise of its municipal authorities, and it also claims
to be one of the driest and sunniest places in England.

Southend Beach

## 10.    The Coast—from Southend to Harwich.

From Southend the coast becomes low and bends
somewhat to the south-east till Shoeburyness is reached.
Shoeburyness, rather more than three miles from Southend,
is a garrison town, and has a " School of Gunnery." Most
of the big guns used in our army are tested here, and long
ranges seawards have been established. A walk along the
beach will bring to view targets of various thickness ; and

after gun-practice the shot are recovered from the sand in which they have been embedded.

The Essex coast now turns to the north-east and we pass a group of islands lying at the mouth of the Crouch. Foulness Point is the extreme point of the largest island, and Holywell Point is on the opposite coast. From this last point to Sales Point the coast runs almost due north, and the district known as Dengie Flats is not without interest. This low marshy coast between the Crouch and the Blackwater is typical of many other parts of the Essex coast. It has been often described, but we select the account given by Mr Rider Haggard, who visited this neighbourhood in 1901. He writes:—

"The view looking over the Dengie Flats and St Peter's Sands from the summit of the earthen bank which keeps out the sea, was very desolate and strange. Behind us lay a vast, drear expanse of land won from the ocean in days bygone, bordered on one side by the Blackwater and on the other by the Crouch Rivers, and saved, none too well, from the mastery of the waves by the sloping earthen bank on which we stood. In front, thousands of acres of grey mud where grew dull, unwholesome-looking grasses. Far, far away on this waste expanse two tiny, moving specks, men engaged in seeking for samphire or some other treasure of the ooze-mud. Then the thin, white lip of the sea, and beyond its sapphire edge in the half-distance the gaunt skeleton of a long-wrecked ship. To the north on the horizon a line of trees : to the west, over the great plain, where stood one or two lonely farms, another line of trees. On the distant deep some sails, and in the middle marsh, a barge gliding up a hidden creek as though she moved across the solid land. Then, spread like a golden garment over the vast expanses of earth and ocean, the flood of sunshine, and in our ears the rush of the north-west gale and the thrilling song of larks hanging high above the yellow, salt-soaked fields."

Not far south of Sales Point are the ruins of St Peter-on-the-Wall, which is supposed to mark the site of a Roman city, which was one of a series built for the defence of the "Saxon shore," and of which very small portions yet remain. About half the site has, however, been destroyed by the inroads of the sea. Rounding Sales Point we enter the large estuary of the Blackwater, and on our way to Maldon we pass the little green island of Osea. Maldon is one of the most ancient towns and boroughs in Essex, and has the unique distinction of a church with a triangular tower.

Following the low marshy coast from Maldon to the north-east, we reach Brightlingsea, which lies on the Colne opposite Mersea Island. This little port is the centre of the yachting interest, and its people are mainly dependent on this pastime. There are always many yachts lying in Brightlingsea Creek, and many more further up the Colne at Wivenhoe. It is a curious fact that Brightlingsea used to be a "member" of the Cinque Ports, and belonged to the borough and port of Sandwich, in Kent. From Brightlingsea the coast bends to the south-east as far as Colne Point, and then makes a curve generally to the north-east, passing Clacton and Frinton, to the Naze.

Clacton has grown rapidly in favour, and is a well-laid out and bracing watering place. From Clacton to the Naze there are cliffs, which are not high, although the sea-views from their summits are fine. Near Clacton is St Osyth's Abbey, with a good entrance gate-house. Walton-on-the-Naze lies on a narrow strip of land,

Clacton-on-Sea

approached by a long winding creek on the north which has its opening in Hamford Water. To the north-east of the old village of Walton the land rises to a kind of head-land, and forms the promontory called the Naze. In Hamford Water are the little islands of Horsea and Holmes, and all this part of Essex bristles with Danish names.

Dovercourt, facing the sea, is half-a-mile from Harwich on the wide estuary of the Orwell and Stour. Harwich has not much accommodation for visitors, but it is nearly always busy with shipping. The town is mean and dull, with an un-English look about it, but there is a pleasant view across the estuary. Harwich is the terminus of the Great Eastern Railway, whose fine boats run daily to the Continent from the quay at Parkeston. The anchorage and shelter at Harwich are so good that it is an important yachting centre. The estuary of the Stour bounds the rest of the Essex coast, and the tide is felt beyond Manningtree, a little junction on the Great Eastern Railway.

## 11. The Coast—the Islands.

A glance at the map of England will show that the insular features of Essex are quite different from those of any other county. Essex is the only county that has so many islands, large and small, quite close to its shores, and it will be well to consider them according to the river mouths in which they are situated. Canvey is in the Thames; an archipelago consisting of Foulness, Wallasea, Potton, Havengore, and New England are at the estuary of the Crouch; Northey and Osea are in the Blackwater; Mersea is at the estuary of the Colne; and Horsea, Holmes, Pewit, and many smaller islands are in Hamford Water, to the south of the estuary of the Stour. All these islands have certain characteristics in common.

With the exception of Mersea, which is a little hilly, they are all very low and marshy, and embankments have been constructed to prevent inundations. Most of them can be reached on foot at low water, and in all cases the surrounding sea is very shallow.

Osea Island

Canvey Island, probably mentioned by Ptolemy as *Kononos*, is most easily approached from Benfleet. If the tide is low, the island may be reached by means of planks and stepping-stones, but if otherwise the ferryboat plies across the narrow channel. Canvey has been termed the most curious place in England, and there is certainly much that strikes the visitor as being quite different from any other part of our country. It is perhaps the nearest approach to Holland that we can imagine. There is a wide stretch of marsh and pasture-land, extending about five miles from east to west, and intersected by drains and cuts. The island was won from the sea by a Dutchman in the early part of the seventeenth century, and there are several Dutch buildings still standing. The village of Canvey is very small, and is partly hidden among trees. There is the little church, faced with boarding, roofed with slates, and surmounted by a bell-cote, and near by are the small houses, some new but others very old. Two round houses built by the Dutch in 1631 still stand, and remind one of the Dutch settlers.

A recent writer says that "Canvey Island lies, a shapeless octopus, right under the high ground of Benfleet and Hadleigh, and stretches out muddy and slimy feelers to touch and dabble in the deep water of the flowing Thames." The views from Canvey are extensive and diversified. Westward, Gravesend is visible, and eastward, Shoebury-ness, while the country around is well cultivated, with pleasant stretches of wood. Hadleigh Castle is within sight, and in the background are the Laindon Hills. Canvey Island no longer suffers from isolation, for it is overrun in the season by visitors from the neighbouring holiday resorts—Southend, Westcliffe, and Leigh.

Foulness Island, a little larger than Canvey, is the chief of the group of islands at the estuary of the Crouch. It is flat and marshy, and the "saltings" in winter are frequented by wildfowl, from which fact the island is said to take its name. Wallasea, Potton, and Havengore lie between its western shores and the mainland, and seawards stretch the Foulness Sands and Maplin Sands,

upon which it is safe to walk at low water. The island
is best reached from Wakering Stairs, some distance north
of Shoeburyness. Across the channel brooms are placed
about every thirty yards, and there are nearly 400 of them.
Every year they are renewed, but it is necessary to repair
many of them at shorter intervals. Across this passage
a pony-cart can be driven, but travellers generally walk
barefoot.

There is another road to Foulness from Burnham.
A walk of four miles along the sea-wall brings one to
Holywell Point, and there an oyster watch-boat can be
obtained to reach the island. Approached from the sea,
the island presents a dreary aspect, the foreshore being
backed by the long brown line of the sea-wall. From
the wall the prospect is not unpleasing, for there is a wide
expanse of pasture, with farmsteads and cottages, a wind-
mill, and the spire of St Mary's church. Foulness, like
Canvey, is a miniature Holland. There are dykes or
sea-walls with the low-lying land behind them ; there are
green pastures ; there are bright-coloured, brown-sailed
barges ; and there are herons, which remind us of the
low country across the North Sea. The island soil was
formerly the richest in the county, and now it produces
good crops of wheat, beans, clover, and white mustard.

Wallasea, the next largest island of this archipelago,
is reached from Burnham, and the ferry is the means of
communication with the other islands. These islands
have similar characteristics to those of Foulness, for there
is the same Dutch landscape of marshes, dykes, windmills,
and cattle.

Mersea is the largest of the many low islands which, separated from the mainland by winding "fleets" and "rays," lie off the Essex coast. The island is hilly, and much more picturesque than either Canvey or Foulness. The value of its position between the Blackwater and the Colne was grasped by the Romans and the Danes, who both settled in it and have left traces of their occupation. The island may be approached from Peldon, and it is connected with the mainland by a causeway, marked out on either side by a row of white stakes. This narrow passage, first constructed by the Romans, is called the *Strood*, and was formerly visible only at low water. It has been recently raised, and is now passable at all times. There was a Roman residence of some importance at West Mersea, and Roman pavements and foundations have been discovered. On the north-east side of the island is Pyfleet Channel, which is famous for its oysters. The marshes around the island are protected from inundations by the sea-wall, a bank of stones and rock about eight feet high forming a path which follows all the windings of the coast. Mersea has two parishes, East Mersea and West Mersea, which are situated at either end of the island. The island farms are divided by thick hedgerows, and the enclosed fields are generally small. The land rises steeply about West Mersea village, whose church is a grey Norman building of flint and stone, with some Roman brickwork.

Finally, a word as to the names of these islands. It will be noticed that many end in *ea* or *ey*. This is the Anglo-Saxon equivalent to the Scandinavian *oe* or island,

and we have the same root again in our Thames word *eyot*, or as it is sometimes spelt, *ait*.

## 12. The Coast. Its Loss and Gain. Its Protection. Sandbanks. Light= houses and Lightships.

In this chapter we are going to conclude our study of the coasts of Essex by considering the loss and gain round its shores, the protection of the coast from the inroads of the sea, and the lighting and buoying of the sandbanks and difficult channels.

The east coast of England has suffered very severely from marine erosion, and every year we lose land equal in size to the island of Heligoland. In Norfolk and Suffolk several places have been submerged, and although Essex has not fared so badly, yet we find that Walton-on-the-Naze and Harwich have decreased in size owing to the demolition of land by the sea. At Clacton-on-Sea the cliffs are crumbling away in many places, and a few years ago several hundred acres of land were ruined by the great flood. Sir Charles Lyell, the great geologist, examined the coast from Harwich southwards, between the years 1829 and 1838, and he had fears that the isthmus on which Harwich stands would, at no remote period, become an island. His fears have been justified to this extent, that between 1824 and 1841 the Essex promontory at Beacon Cliff lost no less than 350 feet.

While the Essex coast is losing in some places, there

has been a corresponding gain in others. Land that had been overwhelmed by the sea has been regained, and specially noticeable is this in the Thames estuary. There land which was once under water has been won from the sea by the unceasing toil of man. It is worthy of remark that this work was accomplished in several instances by foreign engineers, especially those from Holland and Flanders. Vermuyden and Joas Croppenburg are two names that are worthy of remembrance, for they reclaimed marshlands, and rescued and embanked Canvey Island in the Thames.

Canvey Island is very low, and in parts it is actually below the sea-level. In 1622 there was danger that it would become nothing but a sandbank and the owners agreed to give a third of the island to Croppenburg if he would keep out the sea. This Dutch engineer brought over a company of his countrymen, and fought successfully against the sea with the same courage that they displayed in their own country. Croppenburg and many of his friends afterwards settled in Canvey, and built Dutch houses and cottages.

From the earliest times it has been necessary to embank the Thames and the Essex coasts. From Richmond to the sea the Thames is really a channel confined within artificial embankments, and by this means a large extent of fertile land is protected from inundation on both sides of the river. The banks first raised were often of insufficient strength to keep out the water, especially when a strong north-easterly wind was blowing at spring-tide. Down to the end of the seventeenth century, scarcely a year passed without an irruption of the water and a consequent breach in the sea-walls.

Such a breach took place at Dagenham in 1621, which was stopped by Vermuyden. His work, however, was undone in 1707, during a flood and spring-tide driven by a north-easterly wind, when his sluice was damaged and the bank was breached. Many acres of the marsh were utterly lost and washed into the river, there forming a shoal nearly a mile long that blocked up the channel. With every tide 1000 acres of the marshes were drowned, and at low water two great river-like arms were left, stretching across the Dagenham and Havering Levels.

Many efforts were made to repair the damage, and although the owners spent much money their endeavours were long baffled. At length the aid of Parliament was obtained and a grant was voted for this work, which was of national importance. The rift in the wall was fifty feet deep and two hundred feet wide when Captain Perry, who had done some engineering work in Russia, took Dagenham Breach in hand and successfully accomplished his task by an outlay of £25,000. He drove in long piles well fitted to each other, and protected their foundation by throwing in loads of clay. For five years he laboured at this task, and though he succeeded in stopping the breach, he derived no profit from the undertaking. The top of the bank is now fifty feet above low water, and the broad slope on each side shows that the thickness is also considerable. The stopping of Dagenham Breach was a great triumph for Captain Perry, whose engineering ability was afterwards displayed in the Fen district.

All round the Essex coasts it has been found necessary to build sea-walls ; or to construct groynes for the protection of the land. In many respects the county presents a Dutch-like appearance, and the district behind Clacton is known as " Holland."

Let us now turn our attention to the depth of the sea, and the sandbanks off the coast. The sea round our coast is very shallow, and at a distance of 20 or 30 miles from the shore is not deeper than 10 or 12 fathoms, while at the entrance of the Thames it is only 6 or 7 fathoms. Great care has to be exercised in navigating, and vessels generally take a pilot on board to guide them up to London. A glance at the map will show how extensive and numerous are the sandbanks, which in some cases are nearly dry at low water. The largest of these banks are the Maplins, which are covered at high tide with about ten feet of water. Off the Maplins is the measured mile, where vessels of the Royal Navy test their speed. Among other Essex sandbanks are the Foulness Sands, Ray Sands, and Dengie Sands.

It will at once be seen that so difficult a coast as Essex needs to be carefully lighted and buoyed. This work is done by the Elder Brethren of Trinity House, an authority that has " the duty of erecting and maintaining lighthouses and other marks and signs of the sea." They derive their income from light-dues levied on shipping, and are thus able to erect and maintain lighthouses, light-ships, beacons, buoys, etc. They also have the power

to appoint and license pilots, and remove wrecks when
dangerous to navigation. It is worth noting that whereas
a hundred years ago there were only about thirty lighthouses and lightships around the British coast, there are
now nearly 900.

The first Essex light was erected at Harwich in 1666
by Sir William Batten, who in return for his expense in
maintaining this beacon was allowed to levy a tax on
coals and foreign shipping that entered Harwich. He
gained a considerable income by this means, and we find
that this light was thus maintained down to 1863, when
the present lighthouse at Dovercourt was erected. In
1833, the Trinity House built a small lighthouse at Purfleet for experiments with new inventions, and since that
date many lighthouses and buoys have been placed around
the Essex coast. The Nore Lightship marks the entrance
to the Thames, and its white light revolves every half
minute. The Maplin Light is on the south-east part of
the Sand, and its red and white light occults every half
minute.

Among the lightships placed between the sands, the
chief are the Swin Middle, Sunk, Long Sand, and Kentish
Knock. The latter has a white light revolving every
minute, and is connected with the shore by telegraph for
life-saving purposes. Besides the various lighthouses and
lightships on or near the sands, most of the seaports and
coast-towns have their own pier-lights, some of which are
of considerable importance.

In addition to preventing wrecks every effort is made
to rescue life, and for this purpose several of the seaports

The Upper Lighthouse, Harwich

have life-boats. Those stationed at Harwich, Walton-on-Naze, Clacton-on-Sea, and Southend are the most important, and additional help is often rendered by life-boats from the Kentish and Suffolk seaports.

## 13. Climate and Rainfall.

Climate depends on the temperature, the prevailing winds, the dryness or moisture of the air, the character of the soil, and other factors; and we may define the climate of a district as the state of that district with regard to weather throughout the year. In considering the climate of Essex, we must bear in mind that it is a maritime county, having the modifying influence of the sea along a coast-line of more than one hundred miles. It will also be well to remember that it is a large county, and consequently subject to more varieties of climate than we should expect to find in Surrey or in Middlesex. Again, as Essex is further north than Kent, its climate is not quite so warm as the more southerly county.

It is of the greatest importance to have accurate information as to the prevailing winds, the temperature, and the rainfall of a district, for the climate of a county has considerable influence on its productions. Our knowledge of the weather is now much more definite than it was formerly, and every day there appears in our newspapers a great deal of information on this subject. The Meteorological Society in London collects particulars from all parts of the country relating to the temperature of the air, the hours of sunshine, the rainfall, and the direction of the winds. A glance at one of our daily papers will show that the Meteorological Office divides the British Isles into ten districts, and gives the probable weather for the twenty-four hours ending midnight on the day the weather conditions are published. Thus, for October 21, 1907, the following was the forecast for Essex, which is placed in the East England district :—" Southerly and south-easterly winds, strong in places ; weather continuing mild and very changeable, with occasional rain." Warnings are also issued by the same office, so that certain districts may be prepared for the rough weather that is expected. In addition to all this information, some of the newspapers print maps and charts to convey the weather intelligence in a more graphic manner.

There are no fewer than 4000 stations in the British Isles which collect very exact particulars of the rainfall in their districts. These results are arranged in a book called *British Rainfall*, in which we find exactly recorded the number of inches of rain that fell at certain stations. In Essex alone, there are over 80 observers who keep a rain-gauge and enter in a register the daily rainfall. Every year these facts are tabulated for that station, and then forwarded to the editor of *British Rainfall*.

We thus have very definite and full information about the climate of Essex, and are able to compare it with other counties and with the whole of England and Wales. There are so many circumstances that determine the climate of a place, that perhaps it will be well to take one town, Clacton-on-Sea, and consider why its popularity as a seaside resort is so great. First we note that the town faces nearly due south, and is protected by cliffs from northerly winds, and also partly from easterly winds. While the surrounding country is flat and wooded, the town is built on gravel resting entirely on the London Clay. The air is bright, clear, and bracing, and there is much brilliant sunshine. Fogs are very rare, and the rainfall is small. Flowers flourish well, and geraniums often live in the gardens through the winter. The springs are cold, the summers are dry and warm, the autumns are bright, and the winters generally mild. Here one can see that Clacton has many elements in its favour from a climatic standpoint, and we can readily understand why it is such a popular resort.

Of course we cannot go into details about all the towns in Essex, but a few facts will give us a general idea of the climate of the county. First let us take some of the facts relating to the temperature. In 1906, the mean temperature of England was 48·7°, and that of Essex was 49·4°. Thus the Essex temperature is above that of the country as a whole. With regard to the hours of bright sunshine in 1906, we find that, while the average for all England was 1535·5, Essex had an average of 1591·4. Of course some parts of Essex had much more sunshine than this, and to take one instance Clacton had no less than 1945·7 hours of bright sunshine. Thus while places in the neighbourhood of London had dull, cloudy, foggy days, the northern coast of Essex was revelling in sunshine.

The rainfall of England and Wales generally decreases as we travel from west to east. In 1906 the highest

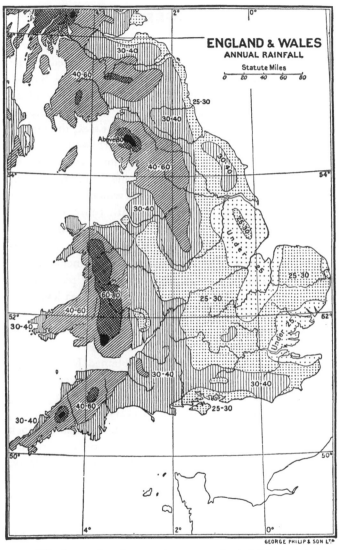

*(The figures give the approximate annual rainfall in inches)*

rainfall occurred at Glaslyn, in the Snowdon district, where no less than 205·3 inches of rain were registered. The lowest rainfall recorded in that year was at Boyton Rectory, in Suffolk, with a total of 19·11 inches. These are both extreme records, but they will serve our purpose, as we find the same law of rainfall in Essex, the highest being generally in the west and north-west, and the lowest in the east. Newport, in the north-west, had 181 rain-days and a rainfall of 26·90 inches, while Clacton in the east had a rainfall of 19·64 inches, which was the lowest recorded in Essex. Shoeburyness with 19·95 inches, and Foulness with 19·99 inches had also a very low rainfall. The highest rainfall in Essex was at Havering-atte-Bowe in the south-west, where it measured 30·38 inches. This is to be explained not only by the westerly position, but also by the place being 342 feet above the sea level, and by various local conditions.

The rainfall records of course vary from year to year, but as they have now been collected for a number of years, the average may easily be obtained for any place which is a station for the collection of weather and rainfall statistics. Thus at Chelmsford, the county town, the average annual rainfall for 30 years was 22·96 inches, and for 1906 it was 24·09, or rather more than one inch above the average. From a further study of rainfall statistics for Essex, it appears that January, February, October, November, and December were the wettest months, and April, July, and August the driest months in 1906.

Now to summarise the main facts with regard to the

climate we may say that Essex is generally dry, the average rainfall being lower than that in any other county. Round the coast and along the Thames the county is subject to cold winds and thick fogs at certain seasons. The table-land of the interior and the elevated portion in

The River Cann at Chelmsford

the north-west are dry and healthy. The severity of the winter is somewhat modified by the proximity to the sea; and the marshlands along the Thames and the sea-coast are less unhealthy than they once were, as they have been drained. When Defoe wrote his *Tour through the Eastern Counties of England* in 1722, he made some very strong statements as to the extreme unhealthiness of the south and south-east of Essex, but two hundred years have worked a great change in this part of the county, and the author of *Robinson Crusoe* would be able to write a very different account of the climate if he were alive to-day.

## 14. People—Race. Dialect. Settlements. Population.

The earliest inhabitants of Essex were probably immigrants from the South and East when the British Isles were still part of Europe. At first these people were ignorant of the use of metals, but as time went on they learnt to work in stone, in bronze, and in iron. There are few written records about these early people till the invasion of the Romans in 55 B.C., when Julius Caesar found the Britons belonging to various races, using different languages, and in different stages of civilisation. The people in Essex were the Trinobantes, a branch of the Kelts, and from various accounts there is evidence that they were skilful in war and were under the leadership of brave chiefs.

As far as Essex was concerned there is no doubt that its conquest by the Romans was complete, and that the natives were Romanised in many ways. After the departure of the Romans at the beginning of the fifth century Essex fell an easy prey to the Teutons from the Continent; and as the Saxons settled in this county a great change was made in the life of the people, who had grown used to the Roman rule. The natives of Essex were gradually driven out, and fresh bands of Saxons settled in the land. The English speech became general, and to show how thorough was the Saxon conquest, it is worth noting that nearly all the places received new names, which are retained to this day. The results of the Saxon conquest were seen in a new language, in new laws, new customs, and a conversion to heathendom.

The Saxons were not allowed to retain their new lands in peace, for after a time there were invasions of the Northmen from Scandinavia, and of Danes from the low-lying lands of Denmark. The latter invaders settled largely in Essex, especially on the sea-coast, which was exactly suited to the seafaring life of these pirates. The Danes gained the upper hand in England, and even Alfred, when he had defeated them, thought it wise to allow them a large portion of our land. From the Thames to the Humber was assigned to them, and in this Danelagh Essex was situated.

A great change was effected in 1066, when the descendants of the Northmen, or Vikings, who had settled in Normandy conquered the land of the English. William the Conqueror imposed his will on the people, and as far as Essex is concerned we can see how thorough his conquest was by referring to the Domesday Book. In that wonderful record, we find that nearly all the Saxon lords and owners of land in Essex were dispossessed and their holdings were given to the Norman lords and their retainers. Norman influence was supreme, and no doubt the "French of Stratford-atte-Bowe," to which Chaucer refers, is only an index to what was going on throughout the county.

The Norman invasion marks an epoch in our history. Since the eleventh century, there has been no hostile invasion of our land, but there have been several periods when foreigners have come over and settled among our people. The Flemings in the reign of Edward III settled in some of the towns in Essex; and the French and Dutch protestants in the reign of Elizabeth and in the time of the Stuarts found a welcome in Colchester, Braintree, Bocking, Halstead, and elsewhere, when they fled from persecution in their own land. In return for the refuge afforded in Essex, these Huguenots and other protestants taught our people many useful handicrafts. The latter portion of the nineteenth century saw an influx of Jews, Poles, and Russians, who, driven from their own country, settled largely in the east of London, and in smaller numbers in West Ham and the adjoining parishes.

From the foregoing remarks it will be gathered that the people of Essex are mainly of Teutonic stock and of English speech. There is an Essex dialect which is the survival of former conditions, and may be heard in the villages far removed from the busy centres, or the railways. Nothing has done more to spoil the Essex speech than the infusion of the "Cockney" element, which has worked sad havoc in the parishes near London.

Having considered the early facts relating to Essex people, we may turn our attention to the people of Essex as we find them to-day. With regard to the population and

Bocking Mill

other statistics we have no certain information till 1801, the year of the union of Great Britain and Ireland. Then the first census was taken, and from that date onwards, there has been a numbering of the people every ten years. The county of Essex was practically the same size in 1801 as it is now, so that we are in a position to compare the various results.

In 1801 the population of Essex was 227,682 and in 1901 it was 1,085,771. This means that the population has increased nearly fivefold in 100 years, a result which is shown by few other counties. During the last ten years the increase has been about 300,000, or 38 per cent., and no doubt the next census will show a very large addition to the number. The density of population in Essex is also striking, for whereas the average population of a square mile in England and Wales is 558, in Essex it is 704.

This enormous increase in the population of Essex has been mainly in the south-west of the county, in the districts of West Ham, East Ham, Walthamstow, and Leyton. In this small area, known as London-over-the-Border, there live nearly three-quarters of the whole population of Essex. The census returns of 1901 show that about 845,000 people live in urban districts, and 240,776 in the rural districts, and that the females exceed the males by 12,847. The bulk of these people live in houses or tenements of which 138,946 contained five or more rooms, and 91,546 had less than five rooms.

The census figures are interesting in many ways. In 1901, there were only 47,159 people in Essex over

65 years of age, and there were no less than 15,479 persons living in workhouses, asylums, and other public institutions. In barracks there were 4766 people, and on H.M's. ships, or on other vessels 3408 people.

Perhaps one of the most interesting tables in the Essex census is that giving the place of birth of the people. We learn that 558,277 were born within the county, 261,533 were born in London, 25,028 in Scotland, Ireland, and Wales, and 1642 in other parts of the British Empire. Persons of foreign birth numbered 9657, and were mainly natives of Germany, Poland, Russia, Italy, and France. With regard to the occupations of the people, the men were chiefly engaged in agriculture, in house-building, as coachmen and servants, in the army, on the railways, or as commercial men and clerks; while the women following occupations were mainly domestic servants, teachers, dressmakers, and milliners.

In 1901 there were 675 blind persons in Essex, and 429 people who were deaf and dumb.

## 15. Agriculture—Main Cultivations. Woodlands. Stock.

Before beginning this chapter it will be best to refer to the diagrams dealing with agriculture, and form in our mind's eye a picture of the relative areas growing the various crops. The two outstanding facts we must first grasp are that Essex has an area of 986,975 acres, and that 793,262 acres of this are under crops and grass. This proportion is a very high one and enables us to realise that Essex is one of the chief agricultural counties of England. Down to the end of the third quarter of the nineteenth century the agriculture of the county was in a most flourishing condition ; the landowners got good rents, the farmers obtained capital prices for their produce, and the labourers were numerous and well paid.

The last quarter of the nineteenth century witnessed a striking change in the prosperity of the county. Rents declined, farmers sold their produce often at a loss, and as a result thousands of acres went out of cultivation, and in common with the rest of the agricultural counties Essex was sorely hit. Of late there has been some improvement, but nothing like a return to the days when wheat fetched 50/- and more per quarter. We shall find considerable interest if we first study the state of agriculture in Essex as it is to-day and then note a few of the crops that are no longer raised in the county.

In a previous chapter we have made some references to the soil of Essex, but here we may remark that the best soil is a friable loam, well suited to the cultivation of cereals and beans. This rich soil runs along the coast, forming a belt of eight or nine miles in depth ; indeed most of the good land in Essex lies low, and the richest pasture is found along the rivers and the marshlands by the borders of the river-mouths and creeks.

We get all our facts about the condition of agriculture from a report issued annually by the Board of Agriculture. That report is of great value and interest, and so we will take it for our guide and discover what it has to tell us with regard to our county. In 1905, as we have said, there were 793,262 acres, or four-fifths of the county, under crops and grass. The corn crops were wheat, barley, oats, rye, beans, and peas, which were grown on 300,664 acres, or three-tenths of the county. Wheat, barley, and oats were the most important crops, and together account for 250,000 acres, or upwards of one-fourth of the entire county. Here it may be mentioned that in the acreage under these three crops Essex is only surpassed by Lincoln and Norfolk.

Beans and peas are most extensively cultivated in Essex, which has 47,323 acres growing these crops against 61,000 acres in Lincolnshire. The green crops, comprising among others potatoes, turnips, mangold, cabbage, and vetches or tares, are grown on 95,845 acres, or one-tenth of the entire county. Permanent pasture in Essex accounts for 285,924 acres, and clover, sainfoin, and grasses under rotation claim 72,279 acres. Very little flax is grown, but small fruit is a rapidly increasing produce and has reached an area of 2018 acres. Essex has a large extent of woodland, which is estimated at 32,415 acres, while the bare fallow land was 36,452 acres.

The cultivation of fruit and vegetables for the early London market is becoming of great importance, and in the neighbourhood of Waltham Cross, Hockley, and many other parts there are glass-houses covering hundreds of acres in extent for the growing of grapes, tomatos, cucumbers, etc. The fruit farms and seed farms are of considerable interest, and are yielding a good return to their owners. Tiptree fruit farm, one of the most famous, covers an area of 300 acres for the production of straw-berries, currants, gooseberries, and raspberries. In the neighbourhood of Witham, there are hundreds of plots devoted to the cultivation of various seeds, such as Shirley poppies, mangold, parsley, swede, nasturtium, and pansy.

Waltham Abbey Church

In the flowering season, this district is quite a blaze of colour and may be seen from the trains travelling towards Colchester. There is one flower, the rose, for which Essex has obtained a world-wide fame, and nurseries particularly at Waltham, Colchester, and Epping grow almost every variety of the national bloom, and the trees and bushes are exported to all parts of the world.

Before we leave the agricultural products of Essex, it will be of interest to recall the fact that in the eighteenth century hops and cherries were both grown to some extent in the county, the former in the north-west, and the latter in the neighbourhood of Burnham and Southminster. There were also some other crops for which Essex was famous—especially the cultivation of coriander, carraway seeds, saffron, mustard, and teasel. The coriander and carraway seeds were used in confectionery, while saffron gave its name to Saffron Walden, owing to its growth in the neighbourhood of that town. Teasel is cultivated for use in the woollen manufacture, and the heads or teasels fixed on frames are applied to the surface of cloth, to raise the nap.

In concluding our review of the Essex Agricultural Report, we must consider the domestic animals. These are classified as horses, cows and other cattle, sheep, and pigs. Of these, in the same year, 1905, sheep were most numerous numbering 212,975. Cows and other cattle numbered 92,888, pigs 82,777, and horses 39,637. Cows are reared to supply milk for London and suburban consumption, and on some of the large estates the yield of milk is very great. Lord Rayleigh farms 10,000 acres of land near Terling, Witham, etc., and has no less than 700 cows of the best breeds for the supply of milk.

At Hadleigh, 36 miles from London, the Salvation

Army have a colony of 1200 acres, which, in return for labour, provides food and lodging to any ablebodied men who are ready to work. All sorts of agricultural work is there practised, besides stock and poultry rearing, and brick-making. On the whole, the colony has been successful, and on derelict land has been the means of giving hundreds of men work which has been remunerative.

## 16. Industries and Manufactures.

The industries and manufactures of a county are largely determined by the physical conditions of the district, and by the facilities offered for the transport of goods. Some of the Essex industries connected with its agriculture, with its fisheries, and with the cultivation of flowers, seeds, fruits, and vegetables have been described in other chapters, and it has been shown that the soil of the county, or the fact of its being a maritime county, has favoured this class of industries. The good railway transport has also encouraged some of them, for it is necessary to convey such articles as milk and garden produce to the markets of the metropolis as rapidly as possible. On the other hand some of the Essex industries have been ruined by the transport facilities. Thus the evaporation of sea-water for its salt, the growth of hops, and the manufacture of cheese have all yielded to the introduction of these articles from more distant counties.

Before we deal with the present industries of the

county, we may glance for a moment at a few of those that once flourished, but are now extinct. Until about sixty or seventy years ago, the making of potash from the ashes of burnt weeds, hedge trimmings, and other vegetable matter was one of the oldest and commonest industries of rural Essex. There are many fields and some farms, which by their names record the fact that a potash factory was on their site; and there is at Radwinter a country inn known as "The Potash." The place of potash for use in soap-making, clothes-washing, etc., has now been taken by soda.

Roman cement was an industry of much importance at Harwich and other coast towns. It was manufactured from septaria, hard stone-like concretions found in the London clay, notably at Harwich and Dovercourt. Roman cement was sometimes known as Parker's cement, from the fact that it was patented by James Parker in 1796. For more than 50 years after that date between 400 and 500 men were employed in the cement trade at Harwich, supplying about two millions of bushels annually. Roman cement has been extinguished by the introduction of Portland cement, which is now made in the south of the county.

Strawplaiting was carried on in the north of the county, and has only declined in quite recent years. It was introduced at Gosfield in 1790, and as a cottage industry it flourished at Castle Hedingham, Halstead, and Braintree. Calico printing was carried on at Waltham Abbey, and silk was manufactured at West Ham. Copper rolling was one of the industries of Walthamstow, and from 1807 to 1845 the British Copper Company had its works in that town. This fact accounts for the name of one of its roads, which is known as Coppermill Lane.

The chief industry of Essex for several centuries was that connected with the manufacture of woollen goods. There is evidence that wool was manufactured in Essex in Roman and Saxon times, and in the Domesday Book

Bocking, Village Street

we have many references to sheep and wool. In 1250 we know that the monastic houses of Essex exported wool to Italy, and there was also a great demand for it in Flanders. At the beginning of the fourteenth century some cloth-workers from Bruges landed at Harwich, and settled at Braintree, Halstead, and Dedham. Edward III gave a great impetus to the wool trade by his encouragement of the Flemings to settle in Essex and teach the people the art of weaving. The chief influx of Flemings was in the reign of Elizabeth, and numbers settled in and around Colchester in 1570, and flourished till about 1748. The clothing towns were Colchester, Braintree, Coggeshall, Bocking, Halstead, and Dedham, and we find that about 60,000 families were employed as spinners, weavers, and combers. The fabrics woven by the Flemings were known as "bays" and "says," and corresponded to our modern baize and serge. Colchester was famous for the "bay and say" trade, while Bocking was noted for its woollen drugget, or baize, which was also known as "Bockings."

We must now devote our attention to a few of the more important of the industries that are carried on at the present time. Romford, Chelmsford, and Colchester are the chief towns engaged in brewing and malting, while the last two and Maldon have some trade in corn-milling. Charcoal-burning was formerly more important than it is to-day, but it is still an industry of some note at Writtle and Hanningfield. The chalk quarries in the north at Saffron Walden, and in the south at Stifford, Grays, and Purfleet, give employment to many hands, and large quantities of this material are used in the manufacture of Portland cement.

Gunpowder was made at Waltham Abbey as far back as 1560, and the works became Government property in 1787. The work is now carried on in 300 separate buildings, which cover 411 acres, and have a water-way of 5 miles. As many as 1200 men are employed, who make annually 2000 tons of cordite, 200 tons of gun-powder, and 150 tons of gun-cotton.

The making of bricks and tiles has been practised in Essex from Roman times, and owing to the abundance of brick earth this industry is one of considerable import-ance at Hedingham, Ilford, Rainham, Dagenham, Grays, Pitsea, Shoebury, and other places in the south. The last census records the fact that 2136 people were engaged in brick and tile making.

Soap and candles are largely manufactured at Stratford and Silvertown. Chemicals, such as camphor, quinine, sulphuric acid, tar, creosote, pitch, naphtha, and turpentine are produced in large quantities at Stratford and Uphall, a part of Ilford. The manufacture of photographic plates is carried on at Ilford, where the works are said to be the largest in the world ; and recently additional works have been opened at Great Warley. Guttapercha and india-rubber goods are made at Silvertown, named after its founder, Mr Silver ; and at the same place there is some sugar-refining.

The manufacture of silk and crape gives employment to 2000 persons at Braintree, Bocking, Halstead, and Earls Colne. The crape made at Braintree is of world-wide fame, and this town had the distinction of making the robe of cloth-of-gold for King Edward VII, and the purple velvet robe for Queen Alexandra, which were worn by them at their coronation. Lace-making is a home industry which employs many cottagers at Cogges-hall, Great Tey, Marks Tey, and Chappel.

Ship-building on the Thames is not so important as it

was once, but the Thames Iron Works on the Essex side
of Bow Creek employ many hundreds of men in their
ship-yards and engineering works.   There is some yacht-
building at Pitsea, Maldon, and Rowhedge, on the Colne.
Chelmsford, Maldon, and Colchester make agricultural
implements, and the county town has also electrical
engineering works, and is developing a trade for building
steam motor-omnibuses.   The Great Eastern Railway
Company have very extensive works at Stratford, where
thousands of men are employed in making steam engines,
railway carriages, and other rolling stock.   The Xylonite
Company's works are at Manningtree and Walthamstow;
and the manufacture of explosives is carried on at Kynoch-
town, in Corringham, and at Stanford le Hope.


## 17.   Fisheries and Fishing Stations.

The fisheries round the coasts of England are very
important, and give employment to many thousands of
people.   Those on the east coast are four times more
productive than those on the west or the south, and this
is chiefly owing to the banks and shoals of the North Sea,
which afford so constant and so large a supply of fish.
As we should expect from the situation and harbours of
Essex, her fisheries, both in the North Sea and in the
estuary of the Thames, are of considerable importance,
and the number of fishermen employed in the season is
very large.

The Essex fisheries are carried on from fourteen or
fifteen stations between Leigh and Harwich, and the

coast is peculiarly favourable for those species of fish which live in a shallow sea with a bottom of sand and mud. The character of the coast line has been fully described in previous chapters, but it will help us if we remember that there is considerable sameness and flatness of aspect throughout.

The sea-fisheries of the county may be thus classified. First there is the North Sea, or deep-sea fishing, in which boats from Brightlingsea are engaged. Secondly, there is off-shore and in-shore fishing, which is practised by the fishermen of Harwich, Tollesbury, and West Mersea. Thirdly, there is estuarine fishing at Maldon, Southend, and Leigh. Fourthly, there is the shell-fish branch, in which Burnham and Wivenhoe are engaged.

There are many methods of catching the fish. Thus for deep-sea fishing the trawl-net and drift-net are used, although on the Dogger Bank some fish, such as cod, are caught singly on long lines and not with nets. When the fish were caught, the old method was to place them in the well of the boat, where they were kept alive by a constant change of water. It is stated that the well-boat was invented at Harwich in 1712, and that by its means fish were delivered in London in a good, and sometimes almost in a living condition. Since the use of steam-carriers and ice, however, these well-boats have lost their former importance, although they are still used in the Dogger Bank fishery.

The other methods of catching fish around the Essex coasts are by means of shrimp-nets, dredge-nets, kettle-nets, and crab and lobster pots. Kettle, or keddell, fishing is used to a limited extent off the south-east coast, at Foulness and Shoeburyness, and is employed chiefly for the capture of the various species of flat fish which frequent the shallow waters covering the sands at high tide. Kettle-nets are about 110 yards long and four feet high. They

The Quay, Mistley

are fixed in position by stakes driven into the ground, and to these the head and ground-ropes are fastened. The nets are in the form of the letter V, and are either set singly, or two or more in a line with the apex, which has a·kind of purse, pointing away from the shore. As the fish follow the rising tide they pass between the nets, and thus find themselves between these and the shore. As the tide falls they are carried into the purse of the net at the apex. The nets are visited at the ebbing of the tide, and the fish are quickly removed.

Seine-netting is rarely used on the Essex coast, but there is a form of trawling for eels on the shores near the mouths of the rivers and on the sand banks of the embouchures. Stow-netting is also carried on extensively, and by its means enormous quantities of sprats are captured in a good season. Another method of fishing peculiar to the Essex coast is known as "petering" or "peter-netting." A peter-net is about 120 feet long and 10 feet wide, with corks on the head-rope and leads on the ground rope, and by this means large numbers of codling, mullet, and other fish are caught. The methods of catching crabs and lobsters are well known to all visitors to our sea-coast towns. The oyster-dredge is like a small trawl, but the mouth is made by a rectangle of iron bands, and the net is usually composed of iron rings linked together.

We do not propose to name all the marine fish caught along the Essex coasts, but we will mention some of the best known and most useful. Flounder, dab, plaice, and sole are common, but halibut, turbot, and brill are rarer. Cod, haddock, and whiting are not numerous, but catches of them are landed at Harwich. The herring is found all round our coast, but there is no special herring fishery, although some are taken in drift-nets in the Blackwater. Sprats are caught in enormous quantities both at the beginning and the end of the year. The importance of sprats is shown by the fact that in Colchester they are known as "weavers' beef."

Among shell-fish, oysters, cockles, and whelks may be mentioned.   Essex oysters are deservedly famous, and those taken from the beds at Burnham and in the Colne fetch a high price.   Colchester celebrates the

Flatford Mill

beginning of the oyster season in October by a municipal oyster-feast, and this shell-fish brings a large amount of money to the town.   The fishermen of Leigh carry on an active trade in shrimps, for which they trawl in the Thames estuary, and also in cockles, which are prepared for the London market.   The "cockling" sheds and the mounds of cockle-shells are familiar to all visitors to Leigh.

With regard to the river fisheries of Essex, we find that the Thames has been poisoned by the discharge of sewage, and that portion of the river belonging to Essex is no longer a salmon river. In the Lea and the Stort, trout, barbel, chub, ruffe or pope, and bleak are caught, besides such well-known fish as perch, pike, and dace. After the Lea, the Colne and the Stour are the most fishful Essex streams, where all the common species may be found. The Cam, in the north-west, has one species, the grayling, which is absent from the other Essex rivers. Excepting the Lea, it may be stated that wherever the trout occurs in any Essex river, it has been of late introduction. Taking the Essex rivers altogether, we find the carp, gudgeon, roach, rudd, dace, chub, minnow, and tench are common, while the salmon, trout, and grayling have almost entirely disappeared.

## 18. History of Essex. I.

When Essex first comes into the light of history, we find that its people, the Trinobantes, were under the leadership of a brave chief, Cassivellaunus, who led a great force of the Britons to oppose Julius Caesar. It would seem that the national defence had been entrusted to Cassivellaunus, who offered Caesar a strong resistance. In the preliminary skirmishes the heavily armed Roman soldiers suffered severely from the dashing onslaught and rapid retreat of the British chariots and cavalry.  As time passed on, the work of Caesar was rendered easier, for domestic discord broke up the British forces, and some of the tribes made their submission to Caesar.

The great struggle between the British and Roman leaders was at Verulamium, the modern St Albans.  This town was filled with a multitude of men and cattle, and defended by forests and marshes.  Caesar attacked this stronghold, and after a brief defence, the natives were defeated, and subsequently Cassivellaunus offered his submission to Caesar.  The Roman general accepted this offer, and having received hostages he departed.  Of Cassivellaunus we have no further information.

After Cassivellaunus, we hear of a chief named Tasciovanus who ruled over the district we now call Middlesex, Herts. and Essex. His capital was Verulamium. We know little or nothing of this chief except from his coins ; and on his death about A.D. 5, he was succeeded by two sons, one of whom Cunobelinus, or Cymbeline, reigned at Camulodunum (Colchester) over the Trinobantes.  Cymbeline has been rendered famous for all time by Shakespeare's play, and his coins are both numerous and well known.  When Cymbeline was reigning at Colchester, Claudius marched against the town with a great army, including, it is said, a number of elephants. The capital of the Trinobantes fell into the hands of the Romans and Claudius took possession of the palace of Cymbeline.  The capture of Colchester involved the downfall of the house of Cymbeline, and Essex fell under the sway of the Romans for nearly 400 years.  Cymbeline had a brave son, Caractacus, who fled to Wales and held out against the Romans until he was captured and sent to the imperial city.

There is one other event in the Roman history of Essex that is worth noting. In A.D. 61 Boadicea, Queen of the Iceni, led an army against the Romans, and severely defeated them. This caused Suetonius, the Roman general, to hasten into Essex in order to avenge the defeat. The two armies met somewhere in the district between London and Colchester. The Britons came on with shouting and singing, while the Romans received them in perfect order and silence, till they were within reach of a javelin's throw. Then at a given signal, they rushed at the Britons, broke their ranks, and pierced through the dense mass. The Romans won a great victory, and Boadicea is said by Tacitus to have ended her life by poison.

The Romans made Camulodunum their chief colony in Essex, which was included in the division of Britain known as Flavia Caesariensis. What the Romans did in Essex, and the results of their conquest, will be told in later chapters.

For some years before the Romans left our country the coasts of Essex and the adjoining counties were often attacked by the Saxons. The Romans had a number of fortresses along the south-east coast to protect it from invasion, and all these fortresses were placed under the "Count of the Saxon Shore." The chief Roman fortress in Essex was Othona, which was probably the little town of Bradwell, facing the North Sea.

When between 400 and 435 the Romans gradually

left our shores, Britain fell an easy prey to the attacks of
the Saxons, Angles, and Jutes. Essex was attacked by the
Saxons, who entered it by the Chelmer and the Stour,
and took possession of the Roman capital, calling it
Colne-ceaster, that is Colchester. The kingdom of the

**Roman Wall at Colchester**

East Saxons took its rise about A.D. 492, and probably
included for some time the modern counties of Essex,
Middlesex, and part of Hertfordshire. Erkenwine was
the first of fifteen kings of Essex, and he began to reign
about A.D. 527, having London for his capital. Essex

had a chequered history in those early days, and it was no uncommon thing for two and even three kings to be reigning at one time. One of the most important events in its history was its conversion to Christianity by Mellitus. This was succeeded by a reversion to heathendom, but at length Cedd, one of the most devoted missionaries of the Early Church, succeeded in converting first the king and then the East Saxons to the true faith.

In A.D. 823, Essex ceased to be a kingdom, and became merged in the larger dominion of Egbert, King of Wessex. During the ninth century the coasts of Essex were ravaged by the Northmen, who found its creeks and openings well suited to their long boats. It was chiefly in the reign of Alfred that the Danes were fighting in Essex, under their famous chief Hasting. The island of Mersea was their camping place, and they appeared at Shoebury, Benfleet, and elsewhere. After a few years of conflict, Alfred made peace with the Danes, who were allowed to settle in the "Danelagh," a district including Essex and reaching from the Thames to the Humber. In 896, a more determined attack was made by Hasting, who took a large fleet of ships up the Lea. Alfred followed his enemy, and besides constructing two forts, one on either side of the river, he diverted the main stream, and so compelled the Danes to leave their boats and flee overland. This was Alfred's last victory, and Hasting troubled England no more. "Thanks be to God," says the old chronicler, "the Danish army had not utterly broken down the English people."

After an interval, the Danes re-appeared, and during

the tenth and early eleventh centuries they gave the people of Essex much trouble. In the reign of Ethelred the Unready, they sailed up the Blackwater to Maldon, where they were met and defeated by the East Saxons under their Alderman, Brihtnoth, who was killed in this great battle. *The Fight of Maldon* is one of our oldest English poems, and from it we learn all about this battle and the brave Brihtnoth. In 1016, another battle was fought at Assandun, on the Crouch, between Edmund Ironside and Canute. Both East Saxons and Danes fought bravely, but owing to a traitor, Eadric, the Danes conquered, and many of the English leaders were slain. King Edmund and Canute the Dane agreed to divide England between them, and as a result of this peace Edmund became King of Wessex, Essex, and East Anglia. Edmund, however, did not live long, and on his death, Canute the Dane reigned in his stead.

After three Danish kings had ruled England, Edward the Confessor was chosen king, and once more the English had one of their own men to reign over them. During the reign of Edward the Confessor, Essex was part of a large district known as Harold's county, and besides Harold, Elfgar, Gyrth, and Leofwine seem to have had their years of government as counts of this territory. Edward the Confessor built himself a palace at Havering-atte-Bower, and Harold founded the church at Waltham. It is generally believed that the body of Harold was buried at Waltham, after the battle of Hastings, and for several centuries the tomb of the last of the English kings was to be seen in this famous church.

## 19. History of Essex. II.

The Norman Conquest was a great event in our history, and affected Essex in many ways. After the battle of Hastings, William lived at Barking while the Tower of London was being built ; but it does not seem that the people of Essex offered any resistance to the Norman Conqueror. We know, however, that William was very severe in his dealings with the Essex lords and landowners, for we find that with few exceptions the manors of this county were given to his own followers. Odo, Bishop of Bayeux, Geoffrey de Mandeville, Eudo Dapifer and many other Normans took the place of the Saxons to such an extent that the natives had little or nothing left to them.

From the Domesday Book we learn much that is of interest relating to Norman England, and perhaps there is no county which receives such full treatment as Essex. So thorough was the survey of it by William's men that we are told "there was not a single mile nor a rood of land, nor was there an ox, or a cow, or a pig passed by that was not set down in the accounts." It is estimated that the population of Essex in William's time was about 80,000, of whom about 14,000 were villeins or slaves. The names of some of the Essex parishes such as Layer-de-la-hay, Stanstead Montfichet, and Tolleshunt D'Arcy are evidently of Norman origin.

We must now pass over the Norman period and consider two events of the greatest importance in the history of Essex. The year 1348 was marked by one of the

greatest plagues that ever visited our land.  The Black
Death, as it was called, swept over Asia and Europe, and
no plague known to history was so destructive of life.
One half of the population certainly perished, and some

Hedingham Castle

think that the number of those who died must be reckoned
at two-thirds.  Many villages in Essex were desolated,
and there were few people left to till the land.  When at
length the plague was stayed, the price of food increased,

and owing to the scarcity of labour, wages rose. Parliament made an attempt to regulate the rate of wages and also the terms of service of the labourers. As a result there was much discontent, and to make matters worse it was decided to levy a poll-tax in order to meet the expenses of the French War.

The peasants revolted in 1381, and in all parts of Eastern England the revolt spread. The Essex rebellion was the work of John Ball, a priest of Colchester, and Jack Straw of Fobbing, who joined their forces to those of the Kentish peasants at Mile End, just over the Essex border. It is a matter of history how King Richard met the rioters, and advised them to disperse. Although the promise of a free pardon was given, the King broke his word, and a commission under Walworth, Lord Mayor of London, was sent into Essex to try the malefactors, nineteen of whom were executed.

The Wars of the Roses did not affect Essex in any marked degree, although some of its nobles lost their lives on the field of battle. The nearest battlefield to Essex was at St Albans, where two battles were fought, in 1455 and 1461.

The end of the Wars of the Roses in 1485 brought the Tudors to the English throne, and began a new epoch in our history. Henry VII curbed the power of the nobles, and on one occasion he visited the Earl of Oxford, a powerful Essex nobleman, at Castle Hedingham. A great company of retainers was drawn up to do honour to the King, and for this the Earl had to pay dearly. "My lord," said Henry, "I thank you for your entertainment, but my attorney must speak with you." The Earl had to pay £15,000, or a sum equal to £180,000, at the present day, for his offence.

The reign of Henry VIII marked the beginning of the Reformation, and all the religious houses in Essex were closed, and as a result there was much suffering and

The Old Siege House at Colchester

discontent. In the time of Edward VI some of the old grammar schools were re-founded, and an impetus was given to education. During the reign of Mary, many Essex people suffered martyrdom at Colchester, Stratford, and Brentwood. When Elizabeth ascended the throne in 1558 a brighter chapter opens in the history of the county, and the year 1588 was rendered memorable by the defeat of the Spanish Armada. It was thought probable that the Spaniards would effect a landing in Essex, either at Harwich or at Tilbury. Both these ports were strongly garrisoned, and at West Tilbury there was an army which comprised 4000 Essex soldiers, clad in blue. The Earl of Leicester, who lived in Essex, had command of these troops, and their courage was strengthened by a visit of the great Queen.

When the Stuarts ruled, Essex had many troubles to face. Its people included a large company of Puritans, some of whom quitted their own land for America, where many settlements were named after Essex towns and villages. In 1642, the Civil War broke out between Charles I and his Parliament, and it would appear that the Essex people mainly sided with Cromwell. However, Colchester declared for King Charles in 1648, and for some weeks it sustained a severe siege by Fairfax. The town was gallantly defended for upwards of eleven weeks by Sir George Lisle and Sir Charles Lucas, but at the end of that time surrendered to Fairfax, who barbarously shot both Lisle and Lucas, whose heroic defence was worthy of better treatment.

The Great Plague of 1665 worked much havoc in Essex, and in Defoe's *Journal of the Plague Year* we have a graphic account of the efforts to prevent the Londoners crossing the Lea into Essex. The south-west of Essex was chiefly affected, and many people died at Waltham Abbey, Epping, Brentwood, Romford, and Barking.

At the beginning of the nineteenth century, Essex

was preparing to resist the invasion of Napoleon.  Martello towers were constructed along the coast at Clacton and

Sir George Lisle

elsewhere, and camps were formed at Danbury, Warley, and Lexden Heath.  Everywhere a patriotic spirit prevailed, and Essex was thoroughly alive to its responsibilities as a maritime county.

Perhaps the two most important events in the nineteenth century as far as Essex is concerned were the passing of the Reform Bill in 1832, and the Repeal of the Corn Laws in 1846. The first resulted in a better representation of the county, and the latter worked a revolution in its agricultural affairs. We cannot close this brief survey of the history of Essex without noting the visit of Queen Victoria to High Beech in 1882, when she declared Epping Forest open to the public for ever.

## 20. Antiquities—Prehistoric, Roman, Saxon.

The earliest history of the people who first dwelt in Essex is not derived from written records, but from the antiquities that have been found in various parts of the county. Antiquaries have divided the earliest periods of our country's history into the Stone Age, the Bronze Age, and the Iron Age. These three periods cover a wide extent of time, and it is not necessary to say how many years are included in each of them, for we cannot be certain when one age ended, and the next began. Following these three periods, we will consider the remains of the British or Keltic, the Roman, and the Saxon times. Antiquities representing all these periods have been found in Essex, but perhaps those of the Roman period are the most interesting and numerous.

Palaeolithic Implement
(*From Kents' Cavern*)

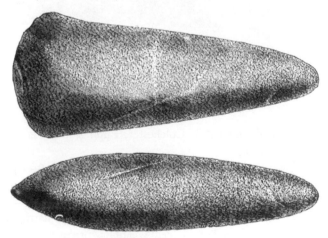

Neolithic Celt of Greenstone
(*From Bridlington, Yorks.*)

The Stone Age men have left some traces of their
handiwork which have been found in the valleys of the
Lea and the Thames.   Flints were used for weapons,
and celts have been found with the bones of the mammoth,
musk ox, reindeer, and Irish elk at Ilford, Walthamstow,
and Walton-on-Naze.   As time passed on the flints were
more carefully worked, and arrow-heads, scrapers, and
even polished celts were fashioned.   Examples of these

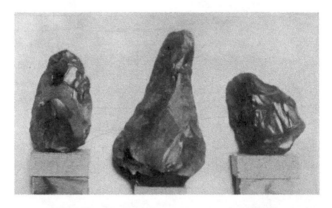

Palaeolithic Implements found at Leyton

have been discovered at Colchester, Epping, and Temple
Mills, Leyton.

The people of the Bronze Age and the Iron Age
were probably of a different race from those of the Stone
Age, for we find that the latter used long barrows for the
burial of their dead, and the former used round barrows.
The Bronze Age has yielded weapons, ornaments, and

pottery, of which good specimens may be seen in Colchester museum. Bronze celts and part of a sword-blade have been found at Shoebury, while bronze spear-heads, leaf-shaped, and having a broad socket, have been unearthed at Plaistow and Walthamstow. In the British Museum there is a fine bronze bowl 16 inches in diameter which came from Walthamstow.

The people of the Iron Age were probably the Kelts or Britons, and specimens of their work in the shape of iron vessels have been discovered at Colchester and Shoebury, while iron celts have been obtained from Walthamstow. Much Keltic pottery has been found at Audley End, and some good urns were discovered at Braintree, in 1903, when a Keltic urn-field, or place of burial, was dug over. British coins, especially those of Tasciovanus and Cunobelinus, are frequently found at Colchester, and on the borders of Hertfordshire.

One of the most interesting discoveries of the British period was made near Walthamstow in 1901, when a British dug-out canoe was found embedded in the gravel at a depth of about 6 feet from the surface. The boat is about 15 feet long and 2 feet 4 inches wide, and was hollowed out of a single piece of oak. Near it were several pieces of early pottery and a well-made iron spear-head.

We will now turn to some very important remains that are generally considered as belonging to the Keltic, or British period. In Epping Forest there are two ancient camps or earthworks, the one, Loughton Camp, covers an area of about 12 acres, and the other, Ambresbury Camp, is generally associated with the name of Boadicea. Both camps have a ditch and a rampart, and each enclosure has the shape of an irregular oval. The camps were recently examined, and among the relics were several flint flakes and axes, and fragments of rough hand-made pottery.

**Group of Samian Ware, Colchester Museum**
(*Red pottery, characteristic of Roman times*)

There are 50 or 60 "dene holes" near Grays, and several have been examined with great care. They consist of narrow shafts or wells about three feet wide and 60 feet deep which are driven into the chalk, leading at the bottom of the shaft into clusters of chambers, usually six in number, and arranged in the form of a double trefoil. It is not certain why these dene holes were dug, but possibly they were formed as refuges, or were used for the storage of grain and food supplies. Far more probably, however, they were quarries for obtaining the flint nodules from the chalk.

Along the east coast of Essex there are numerous "red Hills," which are composed principally of burnt clay and earth mixed with fragments of pottery. It is thought by many antiquaries that they are the waste heaps of ancient potteries.

Some few years ago an interesting discovery of a supposed lake dwelling was made at Braintree, and at a depth of about six feet some British pottery was found. There seems no doubt that this was the site of an early settlement which was the beginning of the present town of Braintree.

There are some remarkable places of burial known as the Bartlow Hills, at Ashdon. These barrows consist of seven tumuli of conical shape, six of them being nearly of the same size, and smaller than the seventh, which is 45 feet high and 147 feet in diameter.

The Roman remains found in Essex are so numerous, that to consider them fully we should want many chapters. We can, however, glance at a few of the more important, and in another chapter some account will be given of the Roman roads in Essex. Of course, Colchester, which was the Roman *Colonia*, is the chief place where all kinds of Roman remains have been discovered. The fine collection of these antiquities in Colchester Museum is one of the most varied and complete in the kingdom, and comprises all kinds of early pottery, vases and urns, and some remarkable pieces of ancient sculpture, known as the Sphinx and the Centurion. At East Ham some lead coffins and a stone sarcophagus were found, and at

Roman Glass Jugs found at Colchester
(*In the British Museum*)

Chigwell were discovered a lead coffin, iron nails, urns, and glass cups. Roman coins and medals have been dug up in immense quantities at Colchester, Leyton, and Saffron Walden. Colchester had the rare privilege of its own mint, and its coins bear the mark C. or CL. There is one other Roman object of special interest, an altar, which was found in 1881, in the Balkern Lane, Colchester. It has an inscription to the mother-goddesses of the heathen world, and there is only one other example of this in Britain.

The most remarkable evidence of the Roman occupation of Colchester are the walls which encircle the town. There is no doubt that a large portion of the walls is the original Roman work, and when we find that their circuit is 3100 yards we realise the importance of the city. Inside the Roman Colonia there were the forum, temples, baths, and theatres, and during recent years many beautiful tessellated pavements have been discovered. If the Roman wall is the most important historical monument of Colchester, the great Roman cemetery that lies on each side of its principal road is of hardly less interest. It has for many years been explored, most valuable discoveries have been made, and Colchester Museum has now a collection of Roman sepulchral remains which is unique in England.

When we come to the Saxon period, or early English times, we have evidence that the Saxon conquest was very thorough. Nearly all the names of places in Essex are of Saxon origin, and that fact alone speaks more eloquently than the finding of many relics beneath the surface. Silver pennies and sceatta of the Saxon period have been found at Bradwell, and other coins at Leyton and Barking.

Some sepulchral pottery was discovered at Broomfield, near Chelmsford, and vases, jewels, gold rings, glass cups, and finger rings have been found in many parts of Essex. The churches that were built by the Saxons in Essex will be considered in another chapter.

## 21. Architecture. (*a*) Ecclesiastical— Churches.

Essex abounds in churches, many of them being of considerable interest. It is true that Essex has no great

Thaxted (from the South)

ecclesiastical centre, such as Canterbury or Rochester in Kent, but it has some churches of great size and beauty, especially those at Saffron Walden, Thaxted, and Coggeshall. With regard to the antiquity of them, we have every reason to believe that there were churches all over Essex at least 200 years before the Conquest, and as we shall see, some of them remain to this day.

In the construction of its churches, Essex has suffered from the absence of building stone. As we have read in earlier chapters Essex is on the London clay, with outcrops of chalk, and for many centuries the greater part of the county was covered by forest. Hence we find that wood, brick, chalk, and flints are the materials generally used in the ancient churches. In parts of Essex there was a plentiful supply of remains of Roman date, and bricks and tiles from Roman camps, villas, and houses were worked into many of the buildings. Where chalk could be obtained, it was used for the interior walls, and sometimes even for pillars and arches. Flints from the chalk quarries being indestructible were very suitable as a facing for the outside walls.

Where the builders could afford stone it was imported from Caen, or Kentish ragstone or Surrey greensand was used, but on the whole very little of these materials was employed. When there came the revival of brick-making in the 14th and 15th centuries, some very fine brick towers were raised, which are yet the glory of several parishes in Essex. There is another material called "pudding stone," which enters largely into the structure of some of the churches. Pudding stone is an aggregation of oblong and rounded pebbles of flint, about the size of almonds, and usually black, embedded in a hard matrix or cement of a light yellowish brown. But of all the materials, timber was undoubtedly the most important, and the early church builders in Essex made the most skilful use of it, not only in the interior work, but also in the outer walls and porches.

There are churches of fine size and proportions in the county, but speaking generally the buildings are small, often consisting of chancel and nave, with only a wooden belfry. Many of the larger and finer churches are in the north, north-west, and middle, while the smaller are in the south. There is little doubt that a beautiful church like that at Coggeshall was built when the woollen trade made its people very rich and prosperous.

Towards the end of the twelfth century the round arches and heavy columns of Norman work began gradually to give place to the pointed arch and lighter style of the first period of Gothic architecture which we know as Early English, conspicuous for its long narrow windows and leading in its turn by a transitional period into the highest development of Gothic—the Decorated period. This, in England, prevailed throughout the greater part of the fourteenth century, and was particularly characterised by its window tracery. The Perpendicular, which, as its name implies, is remarkable for the perpendicular arrangement of the tracery, and also for the flattened arches and the square arrangement of the mouldings over them, was the last of the Gothic styles. It developed gradually from the Decorated towards the end of the fourteenth century and was in use till about the middle of the sixteenth century.

The churches are of all styles ranging from what is

known as Saxon to Perpendicular. It is difficult to say when a church was built, as in many cases the records of its foundation are lost; and sometimes an old church of wood gave place to one built, of more durable materials. Bede tells us that in his day there was not a stone church in all the land, but that the custom was to build them of

St Peter's, Bradwell-on-Sea (from the South)

wood. Hence there is no doubt that the early wooden churches were destroyed by fire, or by some other cause, and the present churches may stand on their sites.

Among the most ancient churches in Essex, and most probably of Saxon date, we may mention St Peter's at Bradwell-on-Sea, St Giles' at Great Maplestead, and Holy

Trinity at Colchester.   Of these three, the first is the
most interesting, and probably the most ancient.   The
ruins of this little church, here figured, occupy the
site of part of the Roman station of Othona.   The
chancel has gone, but the nave, now used as a barn, is
fairly complete.   It is generally agreed that it is the
church mentioned by Bede as having been built by Cedd,
after he had been made Bishop of the East Saxons, in 653.
If this is the case, we have here a building that has stood
through thirteen centuries, and reminds us of the re-
introduction of Christianity into England.

Among churches of the Norman period, St Botolph's
at Colchester is constructed of much Roman brick, and
St Mary's at East Ham is almost unique in having a
double chancel.   Churches of Early English style occur
at Maldon and Great Sampford, while those that deserve
most attention of Perpendicular and Decorated styles
are St Mary's at Thaxted, St Mary's at Saffron Walden,
St Mary's at Dedham, and St Peter's at Coggeshall.   The
first two are among the finest churches of this kind in
England.   Thaxted church dates from a period previous
to 1349, when its construction was stopped by the Black
Death.   The tower and spire, and the fine north porch,
are late fourteenth century work.

The church at Little Maplestead is interesting as
being one of the four round churches in England.   It
consists of a round-ended chancel, with a six-sided western
tower, surrounded by a circular aisle, which was spanned
by arches from the lower arcade.   This peculiar plan was
lue to the fact that the church belonged to the Knights

of the Order of St John of Jerusalem, who settled here
in 1186. The present church dates from the fourteenth
century, but has been much altered and restored.

Another early and most interesting church is that at
Greenstead, near Ongar. It is a mere log-hut, built of
the trunks of trees, and was originally erected as a sort of

Dedham Church

shrine for the reception of the body of St Edmund, when
on its return to Bury St Edmunds, in 1013.

Willingale is the only instance in this county where
there are two churches in the same churchyard, which is
an irregular piece of ground, but nearly equally divided
between the parishes. The west walls of the two

churches are in a line with each other, and the two
buildings are 150 feet apart.

Some of the Essex churches have fine and lofty towers,
especially those at Dedham, Newport, Prittlewell, and
Ingatestone, while stone spires occur only at Thaxted
and Saffron Walden. The tower of All Saints', Maldon,
is triangular, and is probably the only one of this shape in
England. Essex has a few churches with round towers,
but these occur more frequently in Norfolk and Suffolk.
The churches of Pentlow, Lamarsh, Bardfield, Saling,
Birchanger, Broomfield, Great Leighs, and Ockendon
have round towers.

The north porch at Margaretting, the screens at
Stebbing and Great Bardfield, and the octangular font at
South Ockendon are among the best of their kind. There
is not much stained glass in the church windows, but the
monumental brasses are numerous and of considerable
interest.

## 22. Architecture. (*b*) Ecclesiastical— Religious Houses.

In the previous chapter we considered the architecture
of the churches of Essex. We must now turn our
attention to the religious houses which once existed in the
county, but of which we can now see only the ruins.
Before the Reformation in the middle of the sixteenth
century, England was dotted over with abbeys, monasteries
and other religious houses, which were often fine speci-

mens of the architect's skill. In some of our counties, especially in Yorkshire, there yet remain enough of these buildings, either entire or in ruins, to impress upon us their beauty, and to convince us of the large sums of money spent on their construction. While there are no religious houses in Essex that can compare with the beauty of Kirkstall Abbey in Yorkshire, or of Tintern Abbey in Monmouthshire, we shall yet find that many of those in our county were of considerable importance, both from their religious and social influence.

Before dealing with some of the chief religious houses we will consider shortly the material parts of a monastery, and its great officers. In any account of the parts of one of these establishments the church must come first, for it was of necessity the very centre of the regular life. It was generally situated on the north side of the monastic buildings, and its high and massive walls afforded the inmates a good shelter from the rough north winds. In every monastery the cloisters came next in importance, their four walks forming the dwelling-place of the community, and surrounding the cloister-garth. The refectory was the common hall for all conventual meals, and was almost always placed as far as possible from the church. Near to the refectory was the kitchen, which was often of great size, and a small courtyard with the usual offices adjoined it. The chapter-house was on the east side of the cloister, as near the church as possible. Its shape was usually rectangular, and seats were arranged along the walls for the monks. The dormitory contained the cubicles or cells, and every monk had a little chamber to himself. The infirmary, or house for the sick and aged; the guest-house always open to give hospitality to strangers; the parlour, or place of business; the almonry, where the poor could come and beg; the common-room where the monks warmed themselves in winter; and the library, where the books and manuscripts were kept, complete the chief parts of a monastery.

Now with regard to the rulers and great officers of a monastery, the abbot (*i.e.* the father) was supreme, and the entire government depended upon him. The prior, or second superior, was appointed by the abbot, and was concerned with the discipline of the monastery; and the sub-prior was the prior's assistant in the duties of the office. Besides these three great rulers, there were many other officials, each of whom had extensive powers in his own sphere. Thus, among others, there were the chamberlain, who presided over the dormitory, paid the stipends and pensions, and looked after the vestments; the sacristan, who had charge of the church; and the cellarer, who was responsible for the food.

It is generally admitted that most of the religious houses were seats of learning, and the homes of men and women who helped the peasants and promoted the prosperity of the country-side. There were no less than forty-nine religious houses in Essex, and the majority were founded during the two centuries after the Norman Conquest. Many of these abbeys and monasteries were very wealthy, for they possessed hundreds, and, in some cases, thousands of acres of the best land surrounding their houses. It was no doubt with a view to enrich himself and some of his courtiers that Henry VIII laid violent hands first on the lesser, and then on the greater monasteries. From his own point of view, Henry was successful, for we find that the Essex monasteries yielded him a sum of over £6000, which would probably be about £100,000 of our money.

The king's chief agent in Essex was Sir Thomas Audley, and the attack on the lesser monasteries began in 1535 by the dissolution of the Magdalene Hospital, in Colchester. This house, besides affording a home for poor people, had given a daily dole to many more. Its friars were driven out of their home, and were no more seen visiting the sick and poor in their own homes.

In 1537 the dissolution of the greater houses began, and the suppression of St John's Abbey, in Colchester, affords one of the most notable incidents of the dealings with the religious houses. John Beche, the abbot, refused to comply with the Act, and when asked to surrender the abbey to the king's representative, replied, "The King shall never have my house." It was thereupon decided to compass his death, and this was done by treachery. He was asked to a feast, when the magistrates showed him a warrant for his execution, and took him and hanged him without further warning or ceremony, on December 1, 1539. The monks of St John's were turned adrift with small pensions, and the fine buildings were destroyed so that the very site of them is not known, and only the gate-house is now left.

The Priory Church of St Botolph was fortunately preserved for the use of Colchester, but its property was

was raised it was decided that the leaders of the Royalists should be executed. Both Lucas and Lisle were marched to a spot on the north side of Colchester Castle, and there they bravely met their death.

Let us now pass from the men of action to the men of letters, and we shall find it best to consider them under several divisions. Among the famous divines, Samuel Harsnett stands first. For some years he was vicar of Chigwell, and then by successive stages he was Archdeacon of Essex, Bishop of Chichester, and Archbishop of York. He founded the Grammar School at Chigwell, which is one of the best in the county, and at his death in 1531 he bequeathed his valuable library to the Corporation of Colchester. John Rogers, the first martyr in the Marian persecution, was rector of Chigwell. He suffered death at Smithfield in 1555. It is quite remarkable how many eminent divines have been connected with Waltham Abbey. Fuller, the church historian, was curate of Waltham, and here he wrote *The Church History of Britain*. Bishop Hall, the great preacher, was for 22 years curate of Waltham ; and John Foxe, the martyrologist, is said to have written his *Acts and Monuments* in the same town. William Paley, author of the *Evidences of Christianity*, was vicar of Chigwell, and John Strype who wrote much connected with the Reformation period was for 68 years vicar of Leyton. Coming to more recent times, we find that Charles Spurgeon, the celebrated Baptist preacher, was born at Kelvedon, in 1834.

The poets who were born or who lived in Essex form quite a goodly company. George Gascoigne, one of the nave yet remain, and are principally built with Roman bricks, covered with a kind of stucco. St Botolph's Priory is probably unique among Norman churches, and its ruins are certainly beautiful and interesting.

Another of the famous religious houses in Essex was St Osyth's Abbey, near Clacton, whose fine gate-house of rich inlaid flintwork may yet be seen. This monastery is

St Osyth's Abbey near Clacton-on-Sea

said to have been built upon the site of one founded by St Osyth, who was martyred by the Danes in 635. When the house was refounded in 1118, the bones of the saint were enshrined in the new church. The abbey was surrendered in 1539, when it was worth £677, or from £15,000 to £20,000 a year at present value.

The remains of the other religious houses in Essex are

scanty, and not very interesting. There is little left of the great nunnery at Barking, which was the earliest, as it was the most famous and important in England. The abbey at Coggeshall has recently been most carefully restored, and at Leighs the fine gateway remains. The ruins of Beleigh Abbey, near Maldon, are of interest: at Dunmow the church alone is preserved. Of the famous abbey of Waltham Holy Cross, the church remains, and is one of the finest and earliest examples of Norman architecture in England. A religious house existed at Waltham from the days of Canute, but its history is inseparably connected with its foundation by Harold, who is said to have been buried in its church after the battle of Hastings. It was from this abbey that the English took their war-cry of "The Holy Cross!" heard not only on Senlac, but on many another field of battle.

## 23. Architecture. (c) Military—Castles and Moated Houses.

The building of castles is usually associated with the early Norman kings, but many strongholds or castles were built by the Saxons. Perhaps the very earliest idea of these strongholds may be traced in the camps and earthworks which were formed in many parts of Essex, and to which we have referred in a previous chapter. Some church towers were used as places of refuge, and often the old houses were surrounded by moats.

We shall be quite right in saying that by far the larger number of English castles were built during the Norman period. The Norman castles were not all of the same size and importance. Some were royal castles, built for the defence of the country, and placed under the charge of a constable or guardian. Others were built by the Norman barons for the defence of their own possessions, and often became the terror of the country-side.

Most of the great castles were built on the same plan, so that we may as well briefly consider the construction of one of the best type. The area comprising the whole of the castle buildings was within a lofty and thick wall, with towers and bastions, and protected by a moat or ditch. Within this area there was, first, the outer bailey or courtyard, the approach to which was guarded by a towered gateway, with a drawbridge and portcullis. In this bailey were the stables, and a mount of command, and of execution. There was, secondly, the inner bailey or quadrangle, also defended by gateway and towers, and containing the keep, the chapel, and the barracks. The donjon or keep was the real citadel, and was always provided with a well.

As far as Essex is concerned it appears that there were ten or eleven castles, some of which must have been of considerable importance. At Ongar and Pleshey only mounds and ditches remain; at Hadleigh there are fine and extensive ruins; while at Colchester and at Castle Hedingham are grand Norman keeps, ranking among the most interesting and important in England. Let us find what is known about the construction of these fortresses.

It is now generally admitted that Colchester Castle
was the work of William the Conqueror, and no doubt
played its part in the subjugation of Essex. The castle
consists of an enormous Norman keep standing on the
edge of a steep slope, within an irregular enclosure of
earthworks. It is noteworthy as the largest of all the
Norman keeps now remaining in England, the others

Colchester Castle

most nearly approaching it in size being those at London,
Norwich, and Canterbury. It bears a remarkable resem-
blance to the Tower of London, and this is probably due
to the fact that they were designed by the same architect.

There is no doubt that Colchester Castle was built of
the Roman masonry, tiles, and other material from the

ruins of the basilica, forum, and baths of Roman *Colonia*. The area of its ground plan is considerably larger than that of the Tower of London, and, like the latter, it has at the south-east corner a semi-circular projection, forming the apse of the crypt. The upper half of the keep has been destroyed, but it contained a great hall, of which part of a window yet remains. The stairway, sixteen feet in width, the widest in the kingdom, is at the south-west angle of the building. The keep is entered by an arched doorway, which was probably cut through the wall after the tower was built.

Though it was the mightiest Norman castle in England, Colchester Castle never played an important part in our history. It was taken and retaken in the reign of John; suffered little damage in the siege of Colchester in 1648; and was afterwards sold to a man who bought it for the sake of its materials. Fortunately, owing to the hardness of the Norman cement, and the solidity of the masonry, he was baffled in his designs, and was glad to sell the ruins to one who knew how to preserve this fine keep. The castle is now in safe custody, and in the chapel sub-vault is the museum of antiquities found in Colchester and the neighbourhood.

Hedingham Castle, dating from the twelfth century, is by far the most perfect Norman keep in England. The present tower, over 100 feet high, was only part of the castle, which actually consisted of two large baileys separated by a wide and deep ditch. The inner bailey was a mound of more or less artificial character, formed by throwing inwards the material taken out of the deep ditch which surrounded it. The keep is built of rubble-work, cased, or ashlared, as it is termed, with finely squared stone, and its walls are from ten to twelve feet thick. The builder was Alberic de Vere, and it remained in the possession of this important family for nearly five centuries. Hedingham Castle is one of the best preserved Norman towers, for with the exception of the parapets and two of the four corner turrets, the structure is complete from top to bottom.

There is no need to describe the castles at Ongar, Pleshey, Hadleigh, Rayleigh, Saffron Walden and Clavering, for in some instances it is difficult even to trace the plan, and in others the ruins are scanty. We may, however, fitly conclude this chapter by a brief reference to the homestead moats, of which Essex possesses the large number of three or four hundred examples. These water-moated enclosures were usually the site of an ancient manor house or hall, farmhouse or church, but in some cases church, hall, and hamlet were all included within the moat. As a rule the works are rectangular, and the only defence a deep moat, varying from twelve or fifteen feet to sixty feet in width. Some of these moated enclosures may date from the time of the Danes, and were probably constructed as defences against those raiders. There are, however, many houses, especially in the Roothing country, north of Ongar, which were built in the fifteenth and sixteenth centuries, and these are water-guarded in a similar manner.

## 24. Architecture. (d) Domestic—Famous Seats, Manor Houses and Cottages.

The reign of Edward III marks an epoch in the architecture of our country. That great monarch had secured peace and safety for his subjects, and as a result the nobles began to build houses of a more homely character. The necessity for castles, or fortified houses, was passing away, but the barons continued to make their mansions of a very formidable appearance, and obtained leave of the king to wall and embattle them. As time passed on, another great change was made in domestic architecture, and this was brought about mainly as a result of the Wars of the Roses, which broke the power of the great barons. This was specially noticeable in the reign of Henry VII, the first of the Tudors, who made it impossible for any nobleman to exercise undue influence in any district, by restricting the number of retainers he was to accommodate in his mansion.

Hence we may say that in the Tudor period the houses of the great nobles were built less like fortresses, and more as comfortable homes for the owner, his family, and his servants. Fine Tudor houses are still standing in many parts of England, and in our own county there are several good examples. The style of these mansions was largely influenced by the Italian ideas which were then finding favour, and the skill of the architects and the good workmanship of the builders were of great merit. The general plan of the great Tudor architects was to build a big house round a quadrangle having the hall in the middle, and the wings on either side. The characteristics of the building were the quaint gables, large mullioned windows, and grotesque chimneys. The hall was often supported by massive beams of oak ; the lofty rooms had high wainscotting, ornamental plaster ceilings, and beautiful chimney-pieces. Another conspicuous feature in such a building would be the noble oak staircase with its carved balusters and heavy handrail.

The architecture of a county is always influenced by the building materials that are found within its borders. In Essex few buildings are of stone on account of the absence of that material, but brick of varying degrees of merit enters into the construction of many large houses. Wood of course was very abundant, and there is no lack of good timber work both in the mansions and the smaller houses and cottages.

We will now briefly consider a few of the larger mansions of note in Essex, and then devote a little space to the manor houses and cottages. Those that are described must be taken as typical of their class, for it would be impossible even to mention all those of interest, either from their history or their architecture.

The finest and most complete example of a stately house is Audley End, near Saffron Walden. It was built in the reign of James I, and in its original plan it consisted of two large quadrangles, and was approached by a bridge across the Cam. Round the western court were apartments above an open cloister, and steps on its eastern side led to a terrace on which stood the present west front of the house. The inner court was beyond, and the east side was formed by a long and stately gallery. Evelyn, the diarist, describes it as "one of the stateliest palaces of the kingdom," and remarks on the fine decorations and ornaments. The house was so large and the expense of its upkeep so great, that the western quadrangle was pulled

**Audley End House**

down in the eighteenth century. Since then the rest of the building has been carefully restored and it may be said to be one of the best Jacobean houses in England.

At Layer Marney, six miles south-west of Colchester, stand the remains of a fine tower, or gateway. It forms part only of an unfinished mansion of the courtyard type, and was begun about 1520 by Sir Henry Marney, captain of the guard to Henry VIII. The buildings of Layer Marney Tower consist of the gatehouse with towers at the four corners, a wing on the west side now used as a private dwelling, and a range of outbuildings on the west of the house. The walls of the tower are of red brick with chequer-work of black, and there are some

Faulkbourne Hall, near Witham

fine chimney-stacks of twisted patterns. The two octagonal towers are of eight stories, and there is much decorative work in terra-cotta. Indeed the tower is an almost unique example of the early use of terra-cotta in England. Judging from this noble tower, it is quite certain that the projected building must have been a most magnificent design.

Faulkbourne Hall is another splendid example of Essex brickwork of the fifteenth century. It was built round three sides of a small courtyard, which was filled by a staircase at a later date. The most conspicuous features are the lofty tower at one corner, and a turret with a brick spire at another corner.

Gosfield Hall

Horham Hall was the work of Sir John Cutts in 1510, or thereabouts, although there are traces of an older building. There is a moat on three sides of the house, and the great hall is very fine, with its huge fire-place, old ceiling, and stained glass. At the south-west corner of the house there is a fine chimney-stack; and the prospect tower at the north-east angle was added in the sixteenth century, for watching deer-drives in the park.

The splendid mansion of Gosfield Hall, about two miles

south-west of Halstead, was originally a brick house of the reign of Henry VII. It was built round a quadrangle, into which all the windows opened, leaving the exterior face a dead wall up to the first story for the sake of defence. The west side remains almost in its original state, and contains on the first floor a fine gallery, 106 feet long by 12 feet wide.

New Hall near Boreham

New Hall, not far from Boreham, is a fine red brick building of Tudor age and architecture, with bay windows and pillared chimneys. It belonged successively to the Earl of Ormond, grandfather of Anne Boleyn, George Villiers Duke of Buckingham, Oliver Cromwell, and other great personages, and it is thus of interest owing to the great names of its many owners and for its numerous royal visitors, among whom was Henry VIII.

Moyns Park, in the extreme north-west corner of the county, is an Elizabethan mansion and a good example of the ornamental style of that period, with its massive bay windows rising up to the full height of the house, its quaint gables, and its clustered chimneys. Like many old manor houses, Moyns Park has its moat, over which access is gained by a bridge.

Moyns Park

The manor houses of Essex are very numerous, for many parishes have two or three of them. They are generally called " halls," and good examples of this type of architecture are to be found in most parts of the county. The pretty little village of Feering has a timber-built manor house known as Feeringbury which still retains its moat ; and there is an old manor house at Rayne, not far

from Leighs, which was the dwelling-place of a powerful
Essex family, the Capells, who were the ancestors of the
Earl of Essex.    Rochford Hall, once a splendid mansion,
and the reputed birthplace of Anne Boleyn, is now but
a shadow of its former magnificence.    The building is of
red brick, covered with plaster, and the most noteworthy
features are the gables and the fine tall chimneys of
ornamental brickwork.

Many really good houses were built when the clothing
trade was making Essex tradesmen very wealthy.    There
is a remarkable house at Coggeshall which was the work
of Thomas Paycock, who died in 1580.    The building,
known as Paycock's House, is constructed throughout of
timber, and has much fine carving on the ceiling beams
and other parts.

The cottages of Essex are not so picturesque as those
of Kent or Surrey, but many villages of our county have
good examples of cottage architecture.    The village streets
of Messing, Bocking, Dedham, and Chigwell have really
picturesque houses and cottages with old gabled fronts.
They are generally simple in design, but very pleasing to
the eye.    The village of Ford Street, a little hamlet not
far rom Colchester, is a typical example of old English
rustic architecture, and makes a very picturesque view.

While we are considering the subject of picturesque
houses, we may mention the almshouses at Audley End,
a delightful group of buildings in red brick, with irregular
gables and chimneys ; and the row of almshouses, dating
from 1527, in Walthamstow churchyard.    They too are
of red brick, roofed with red tiles, and so mellowed by

age that they present a fine picture of what could be done in an age when men used other materials than yellow bricks and blue slates. The Essex parishes in the vicinity of London have endless rows of commonplace houses, all of the same type, and without any pretensions to beauty. The chief aim of the modern builder is to get as many houses as possible on a given area, whereas the builder in Tudor or Stuart times endeavoured to obtain both variety and beauty.

## 25. Communications—Past and Present —Roads and Railways.

There is no doubt that the Trinobantes who lived in Essex when Caesar invaded our land made paths or trackways through the primeval forest which then covered their territory. The trackways are in many instances the origin of parish roads, for they led from one settlement to another. In some localities where high roads have been constructed portions of the British trackways remain ; and when they have not been gravelled by the local authorities they are known as grass roads. British trackways have been traced in many parts of Essex, but as they have long since ceased to be of importance as main roads, we only refer to them as they were prior to the fine roads made by the Romans.

In our county the Roman roads are of the greatest importance, and give access to all parts of Essex. The main object of the Romans in making these roads was undoubtedly military, and the line of communications between their various forts and stations was carefully guarded. These roads were not only used as great channels of communication, but they also served as the limits of the various divisions of their conquests, while the boundaries were marked by mounds, stones, or trees.

The chief Roman highways are known to us for the most part by the names given to them by our Anglo-Saxon forefathers. Most of the Roman roads converge on London, and coincide in a remarkable manner with our modern railroad communication. Thus in Essex, Icknield Street, corresponding to our Great Eastern Railway, ran from London to Colchester, or Camulodunum, as it was called in those days. This was the main road, and other roads led from Colchester through the north-western parts of the county to the adjoining counties. The importance of the Roman road from London to Colchester is quite evident when we remember that upon it all the principal Roman towns were situated. After a time this main road was extended to Harwich and Ipswich. The great road from London to Norfolk passed through Leyton and other parishes to Cambridge. Another Roman road of importance was known as Stone Street, and led from Colchester to Dunmow ; and there is evidence that a Roman road connected Tilbury on the Thames with the main highway at Brentwood.

We must now leave the times of the Roman and come down to a more recent period. Even as late as the seventeenth and eighteenth centuries the roads of Essex were in a deplorable condition, and it was almost impossible to travel with speed or comfort in any direction. Everything was carried on horseback, and sometimes on bullockback. Corn, coal, wool, iron, and other articles were carried on pack-horses. The roads were mere tracks, and when the old tracks became dangerous through depth of mud, new tracks were struck out across the adjoining fields. Guides were employed to keep travellers out of the mud, which was often very thick.

The Essex high roads were also beset by highwaymen, and the traveller passed along bridleways through fields, where gibbets often warned him of his perils. Pepys, the diarist, gives us many details of the difficulties he encountered when travelling in Essex, and at a later date we know the terror that Dick Turpin was to Essex folk.

In the middle of the eighteenth century some attempt was made to improve the roads and make them suitable for carts, waggons, and carriages. The stone was put on, and the roads were left to manage themselves. Even near London, the Essex roads were often impassable gulfs of mud, and it took longer for a coach to reach a place twenty miles out of London than it now does to reach York or Manchester.

We can get an excellent idea of what the Essex roads were like in the eighteenth century by referring to the travels of Arthur Young. In 1767 he was in Essex, and thus describes the road from Tilbury Ferry to Billericay : "Of all the roads that ever disgraced this Kingdom in the wry ages of barbarism none ever equalled that from Billericay to the 'King's Head' at Tilbury. It is for 12 miles so narrow that a mouse cannot pass by any wagon...the ruts are of an incredible depth...the trees everywhere overgrow the road so that it is totally impervious to the sun except in a few places.... I must not forget eternally meeting with chalk wagons, themselves frequently stuck fast till a collection of them are in the same situation that 20 or 30 horses may be tacked to each to draw them out one by one."

A change for the better was made in the condition of the Essex roads towards the end of the eighteenth century, and many Acts of Parliament were passed authorising the construction of new roads and bridges. In the early years of the nineteenth century a great improvement had taken place, and such men as Macadam were appointed to superintend the roads. This was notably the case on the Epping Road, and the result was seen in the numerous coaches that were running between the chief Essex market-towns and London.

At the present time the Essex roads are in an excellent condition. The most important highway enters Essex at Stratford and proceeds due north-east to Colchester, and thence to Ipswich. The second chief line of road leaves the main road at Chelmsford, and proceeds through Braintree, Halstead, and Sudbury to Bury St Edmunds. The third, or Newmarket Road, leaves the main road at Stratford, and proceeds in a north-westerly direction through Epping, Harlow, and Saffron Walden to New-market.

Not only did the eighteenth century witness a great improvement in the condition of Essex roads, but the county shared with the rest of England in the new spirit of commercial activity. Canals were made, bridges were built, and above all, the steam-engine was invented. In the year 1797 the Chelmer was made navigable by means of locks, enabling barges of 30 tons to convey goods as far as Maldon. Watt's steam-engine of the eighteenth century was to be superseded by Stephenson's locomotive of the nineteenth century, and Essex was soon to have its railways for the transit of passengers and goods. The present Great Eastern Railway is the outcome of the Eastern Counties Railway, which was originated in 1836 and opened in 1839. The Eastern Counties had its terminus at Shoreditch, and was designed to supply a road from London to Colchester. With it was associated the Northern and Eastern Railway, which extended from Stratford towards Cambridge. In 1840 this line was opened to Broxbourne, and shortly afterwards by way of Bishop Stortford to Newport in Essex.

By the year 1862 there were many railway lines in the eastern counties, and then it was deemed advisable to amalgamate them as the Great Eastern Railway, which since that date has increased its mileage and reputation. The present London terminus of the Great Eastern Railway is the great station at Liverpool Street, and in Essex alone this company controls more than 200 miles of railway. Besides the Great Eastern Railway, Essex is served by the London, Tilbury, and Southend Railway in the south, and the Colne Valley and Halstead Railway in the north.

## 26.  Administration and Divisions— Ancient and Modern.

In order to get a just idea of the present administration of our county we must remember that many of our institutions can be traced back for a thousand years or more.  Thus it is our boast as Englishmen that while continental nations have changed their institutions and mode of government by revolutions, we have made our changes by gradual steps, or by a process of evolution.

In Saxon times the government of each county was partly central and partly local.  The central administration was from the county town, and the local administration was carried on in the hundreds and parishes.  The chief court of the county of Essex in the earliest times was the Shire-moot, which met twice a year, and its two chief officers were the Ealdorman and the Sheriff, the last of whom was appointed by the King.  The Shire-moot may be looked upon as the later form of the older folk-moot, and this assembly was held often in the open air on a haunted mound, or round some aged oak of sacred memories.  To the Shire-moot were sent representatives of each rural township and of each hundred.  The Sheriff (Shire-reeve or Shire-steward) published the royal writs, assessed the taxation of each district, and listened to appeals for justice.

In Saxon times, each county was divided into Hundreds, or Lathes, or Wapentakes.  Essex was divided into nineteen hundreds and one Royal Liberty.  Each hundred probably consisted at first of one hundred free families, and had its own court, the hundred-court, which met every month for business. Each hundred was sub-divided into townships, or parishes, as we now call them, and every township had its own *gemot*, or assembly, where every free-man could appear. This gemot, or town-moot, made laws for the township, and appointed officers to enforce its by-laws, or laws of the *by* or town. The town-moot was held whenever necessary, and its chief officer, the reeve, acted as president.

There was thus a kind of triple arrangement for the government of each county ; but besides these courts of the shire, the hundred, and the township, there were also courts of the manor, as the separate holdings of land were called. The manors varied in extent, being sometimes as large as the township, while in many cases they were smaller. The manors were held by their owners, the lords of the manor, on various conditions. For instance, the lords had to render service, or homage, to the King and were allowed to sub-let their manors. The manors had their own courts, such as court-leet, court-baron, and customary court. In these courts, the lord and his tenants met, and settled the affairs belonging to the manor, such as those relating to the common fields, the right of en-closure, and the holding of fairs and markets. These manor-courts are still held in many parts of Essex, and although they have lost much of their original importance, it is interesting to remember that in them we have sur-vivals of the work of our forefathers more than one thousand years ago.

Having glanced at the early administration of Essex, we can now consider its present form of government. The chief officers in the county are the Lord-Lieutenant and the High Sheriff. The former is generally a nobleman or a large landowner, and is appointed by the Crown. The Sheriff is chosen every year on the "morrow of St Martin's Day," November 12th. The County Council is now the central authority and conducts the main business of the county. It was constituted in 1888 and holds its chief meetings at the Shire Hall, Chelmsford. The members of the Essex County Council consist of 24 Alderman and 79 Councillors, the latter being elected by the ratepayers while the former are co-opted.

For local government in towns and parishes, an Act was passed in 1894, when new names were given to the various bodies which had previously been known as vestries, local boards, highway boards, etc. In the towns and larger parishes, the chief governing bodies are now known as urban district councils, of which there are 33 in Essex. Those smaller parishes with a population of over 300 have a parish council, and those with a population under 300 have a parish meeting. The district councils and the parish councils thus represent the old town-moots, and the members are elected by the people to manage the affairs of their locality.

Some of the older and larger towns in Essex have a different form of government. These towns are called Municipal Boroughs, and are as follows :—West Ham, East Ham, Colchester, Maldon, Saffron Walden, Chelmsford, Harwich, and Southend. Each borough is governed

The Town Hall, Colchester

by a Corporation or Town Council, consisting of Mayor, Aldermen, and Councillors.

Essex has also 16 Poor Law Unions, each of which has a Board of Guardians, whose duty it is to manage the workhouses, and appoint various officers to carry on the work of relieving the poor and aged.

For the administration of justice, Essex has one Quarter Sessions at Chelmsford, and 22 Petty Sessional Divisions, each having magistrates or justices of the peace, whose duty it is to try minor cases and award punishment.

Reference was made in the early part of this chapter to the Royal Liberty of Havering-atte-Bower. For a long time it had its own courts of justice, but they were abolished in 1892.

We now pass to ecclesiastical administration, and we find that the mode of government for the Church is much the same to-day as it was one thousand years ago. The Church existed before the State, and had its own dioceses and courts. The northern dioceses were under the care of the Archbishop of York, and the southern dioceses under the authority of the Archbishop of Canterbury. Essex then belonged to the diocese of London, and so continued till the county was transferred to the see of Rochester in 1836, and then to St Albans in 1876. It is now in contemplation to constitute Essex into a separate see having its own bishop. The diocese is divided into archdeaconries, rural-deaneries, and ecclesiastical parishes. At one time, the ecclesiastical and civil parish were one and the same, but now there are many more of the former than of the latter, which number 399, and parts of others in adjoining counties.

For purposes connected with education, the Essex County Council has appointed a County Education Committee, which has control of secondary and elementary education in the greater part of Essex. There are separate education committees for 11 of the largest towns and parishes.

Lastly we come to the parliamentary representation of the county of Essex. For this purpose there are eight divisions, each returning one member—the N. or Saffron Walden, the N.E. or Harwich, the E. or Maldon, the W. or Epping, the S.W. or Walthamstow, the S. or Romford, the Mid or Chelmsford, and the South-eastern. There are two boroughs, Colchester and West Ham, the former being represented by one member, and the latter by two members in the House of Commons.

## 27.   The Roll of Honour of the County.

We shall now find it both interesting and instructive to take a brief survey of the worthies of Essex. Its Roll of Honour is a long one, including as it does the names of many who have shed lustre not only on this county but also on Britain. It is well for us to associate the names of great men with the localities where they lived, as it gives an added interest to know where the worthies of a county passed their lives and to visit the places which they have rendered famous, either by their writings or by their deeds.

Let us first glance at some of the royal personages who lived in Essex, or who frequently visited it. As we might suppose from the fact of Essex being one of the metropolitan counties many of our sovereigns are associated with it. Edward the Confessor built a palace at Havering-atte-Bower, where he passed much of his time. This is no doubt the reason why the district formerly known as the Liberty of Havering had peculiar privileges. Harold, the last of the Saxon kings, had large possessions in Essex and rebuilt the church at Waltham, where also he founded a college. Harold visited Waltham Church on his way to meet William at Hastings, and after that great battle was buried in the church he had rebuilt. The Conqueror retired to Barking, which was then a great town with a famous abbey, and there he lived while he was building the Tower of London. Queen Matilda, wife of Henry I, often visited Barking Abbey, and it was she who built a bridge over the River Lea, so that Barking might be reached with safety. This bridge was called Bow Bridge, because it was arched like a bow. It was famous, too, because it was the first stone bridge constructed in England.

Epping Forest proved a great attraction to those of our monarchs who loved hunting. Edward IV hunted with the chief citizens of London in this forest. Queen Elizabeth was very fond of hunting and frequently enjoyed the sport it afforded. There still remains at Chingford a building called Queen Elizabeth's Hunting Lodge, which commemorates the great Tudor Queen. Not only did Elizabeth visit the Forest, but she also made several "Progresses" through the county, visiting her nobles at their great houses. Thus we have records of the receptions she met with at Wanstead, Havering, Colchester, Harwich, and Saffron Walden. Queen Victoria visited Essex in

Thomas Cromwell, Earl of Essex

1882, when she opened Epping Forest to the public for ever.

We may now notice a few of the statesmen who lived

in Essex. Lord Audley, who worked his way to the Lord Chancellorship, was one of Henry VIII's most willing agents in closing the monasteries and appropriating their revenues. He came in for a large share of the abbey lands in Essex, and with the proceeds built himself a lordly house at Audley End. Thomas Cromwell, whose career covered about the same period as that of Lord Audley, was another of Henry's trusted agents in the work of spoliation. He was named "the hammer of the monks," and for his service to Henry he became Earl of Essex. Passing from the sixteenth century to the nineteenth, we come to a statesman, Benjamin Disraeli, afterwards Lord Beaconsfield, who was educated in Essex. He spent some of his school days at the Rev. E. Cogan's house at Higham Hill, Walthamstow, and there he came in contact with many clever boys who rose to fame in after life. The house still stands at Higham Hill, and the schoolroom in which the young Disraeli studied may yet be seen.

Among the great Essex soldiers we shall select Sir John Hawkwood, Sir Charles Lucas, and Sir George Lisle. Hawkwood was born at Sible Hedingham about 1320, and after serving under Edward III, he became illustrious as the leader of a band known as the "White Company," which fought against Milan. He had a most adventurous career in Italy, where he won a great name for himself. He died at Florence in 1393, but his body was buried in his native place. Sir Charles Lucas and Sir George Lisle were the defenders of Colchester when it was besieged by the Parliamentary army under Fairfax. When the siege

Sir Charles Lucas

was raised it was decided that the leaders of the Royalists should be executed. Both Lucas and Lisle were marched to a spot on the north side of Colchester Castle, and there they bravely met their death.

Let us now pass from the men of action to the men of letters, and we shall find it best to consider them under several divisions. Among the famous divines, Samuel Harsnett stands first. For some years he was vicar of Chigwell, and then by successive stages he was Archdeacon of Essex, Bishop of Chichester, and Archbishop of York. He founded the Grammar School at Chigwell, which is one of the best in the county, and at his death in 1531 he bequeathed his valuable library to the Corporation of Colchester. John Rogers, the first martyr in the Marian persecution, was rector of Chigwell. He suffered death at Smithfield in 1555. It is quite remarkable how many eminent divines have been connected with Waltham Abbey. Fuller, the church historian, was curate of Waltham, and here he wrote *The Church History of Britain*. Bishop Hall, the great preacher, was for 22 years curate of Waltham ; and John Foxe, the martyrologist, is said to have written his *Acts and Monuments* in the same town. William Paley, author of the *Evidences of Christianity*, was vicar of Chigwell, and John Strype who wrote much connected with the Reformation period was for 68 years vicar of Leyton. Coming to more recent times, we find that Charles Spurgeon, the celebrated Baptist preacher, was born at Kelvedon, in 1834.

The poets who were born or who lived in Essex form quite a goodly company. George Gascoigne, one of the most famous Elizabethan poets, lived in Walthamstow, and some of his poems are written from " my poore house at Waltamstow." Both the *Steel Glass* and *The Complaint of Philomene* owe much of their local colouring from

GEORGE GASCOIGNE.

Gascoigne's knowledge of the Forest. Thomas Lodge, another Elizabethan poet, wrote his *Wit's Misery* at Leyton. Thomas Tusser was born at Rivenhall in 1515, and after being educated at Eton and Cambridge he spent some years at court and then turned his attention to farming. He wrote *A Hundred Points of Good Husbandry*, which he expanded some years later into *Five Hundred Points of Good Husbandry*. Francis Quarles, author of *Emblems*, and other religious poems, was born at Romford in 1592. George Herbert, who was a contemporary of Quarles and whose life was written by Izaak Walton, wrote *The Temple* and other poems of an ardently devotional character which will always render them imperishable. He lived with his brother Sir Henry Herbert for some years at Woodford.

Coming down to the nineteenth century, Tennyson lived at Beech Hill House, Epping, where he wrote *The Talking Oak* and *Locksley Hall*. Coventry Patmore was born at Woodford in 1823, and although his rank as a poet is not high, he is remembered by *The Angel in the House*. Thomas Hood, author of *The Song of the Shirt*, lived at Lake House, Wanstead, where he died in 1845. His novel *Tylney Hall* was suggested by the story of the famous Wanstead House, and one of his poems *The Epping Hunt* was especially characteristic of this writer's humour.

William Morris, the greatest of the Essex-born poets, first saw the light of day at Walthamstow in 1834. In that parish and the neighbouring Woodford, Morris passed his early years and received his education. Epping Forest and the features of the Essex landscape made a deep impression on him. He describes the "wide green sea of the Essex marshland," "the dense hornbeam thickets, the biggest hornbeam wood in these islands, and I suppose in the world," and there is no doubt that the author of *The Earthly Paradise* and *The Life and Death of Jason* owed much to the happy years he passed in his homes on the borders of Epping Forest. Elm House, where Morris was born, no longer stands, but Water House, Walthamstow, one of his residences, a square, heavy Georgian building of yellow brick, is carefully preserved by the District Council.

Philip Morant, the learned historian of Essex, was vicar of Aldham towards the end of the eighteenth century. His *History and Antiquities of the County of Essex* is the standard work, and in many respects is not likely to be superseded. William Stubbs, who became Bishop of Oxford, was for some years vicar of Navestock, where he gathered his material for *The Constitutional History of England*, a work of the greatest learning.

Leaving the historians, we will glance at some of the men of letters who have lived in Essex. John Locke stands first both in time and importance as a philosopher. For many years he lived at High Laver, where he was buried in 1704. It was there that he wrote *Letters on Toleration*, which is worthy to rank with his greater books *Essay concerning the Human Understanding* and *Thoughts on Education*. Sydney Smith, the witty canon of St Paul's, was born at Woodford in 1771. Besides works of a religious character, he is famous for essays in the *Edinburgh Review*. Charles Dickens, in many ways the greatest novelist of the Victorian era, often visited Essex. He was particularly fond of Chigwell, and one of his novels,

*Barnaby Rudge*, is connected with that pretty village. Dickens writes of Chigwell as " the greatest place in the world...Such a del cious  ld inn facing the church...Such a lovely view...Such forest s enery...Such an out-of-the-way rural place...Such a station ! " Besant's novel, *All in a Garden Fair* gains much of its interest from the picturesque descriptions of Hainault Forest, which the novelist knew so well.

Samuel Purchas was born at Thaxted in 1575. He devoted much of his life to the great collection of voyages and travels known as *Purchas his Pilgrims*, and *Purchas his Pilgrimage*. Samuel Pepys, the celebrated diarist, had many friends in Essex whom he often visited, and described in the *Diary*. Daniel Defoe, the author of *Robinson Crusoe*, knew Essex well. He had property in Colchester, and brick and tile works at Tilbury, where he had a house by the riverside. His *Journal of the Plague Year* gives a vivid account of the plague-stricken Essex parishes, and his *Tour in the Eastern Counties* is of great value as it portrays the state of Essex in the early eighteenth century.

Among English men of science, Essex can claim three of the first rank. William Gilbert, physician to Queen Elizabeth and James I, was born at Colchester in 1540. He declared the earth to be a magnet in his *De Magneto*, the first great scientific book published in England. He died in 1603 in his native town. William Harvey lived at Hempstead, where he died and was buried in 1657. Harvey made the great discovery of the circulation of the blood and wrote several works of great value. One of

William Harvey, M.D.

our historians asserts that "only two discoveries of any great value came from English research before the Restoration," and these were made by Gilbert and Harvey. John Ray, the son of a blacksmith, was born at Black Notley in 1628. He was a great naturalist, and the first

Constable's Birthplace, Flatford

to raise zoology to the rank of a science. Modern botany began with his *History of Plants*. His varied labours have justly caused him to be regarded as the father of natural history in this country, and as a botanist he has won the highest commendation from his successors.

Before we close this chapter we must briefly refer to three persons—William Penn, Mrs Fry, and John Constable—who were intimately connected with Essex. William Penn, the founder of Pennsylvania, lived at Wanstead, and was educated at Chigwell Grammar School. After a busy life in America he returned to England and died in 1718. Mrs Elizabeth Fry spent her married life in Essex at Plashet, Plaistow, and Barking where she was buried in 1845. Mrs Fry was a prison reformer and interested herself in the welfare of the lowest classes. John Constable, one of England's greatest landscape artists, was a native of Dedham, and received his early education at the school in that town. He painted in his own style quiet English landscapes, such as he was familiar with on the banks of the Stour. For a long time he worked without recognition in England, but his work is now highly valued.

## 28. THE CHIEF TOWNS AND VILLAGES OF ESSEX.

*(The figures in brackets after each name give the population in 1901, and those at the end of the sections give the references to the text.)*

**Ardleigh** (1426) is a pleasant rural village four and a half miles north-east of Colchester. Its church has a lofty tower of flint and brick, which serves as a landmark.

**Aveley** (1060), a large village, is about two miles from Purfleet. The church is of Norman and Early English origin, and has some interesting brasses. In this parish stands *Belhus*, an excellent example of a Tudor mansion, surrounded by a park of great extent, with fine views across the Thames.

**Baddow, Great** (2308) is one of the pleasantest villages in Essex. Its church, with a large and handsome tower, dates from the fourteenth century.

**Baddow, Little** (510) is situated five miles north-east from Chelmsford. Its church is a small and ancient structure; and in the neighbourhood are some very picturesque spots among the extensive woods.

**Barking** (21,547), a large market town on the River Roding, was once famous for a large fishing fleet. A nunnery was founded here in the seventh century, and became of considerable importance, having many noble and even royal abbesses. Little remains of this abbey, except the Firebell Gate. Eastbury House, an old Tudor mansion, has been connected with the Gunpowder Plot. The main outfall of the London sewage is in this parish, and there are extensive market gardens. (pp. 6, 39, 94, 103, 140, 151.)

**Bentley, Great** (1156), eight miles south-east of Colchester, is a large pleasant village with an extensive green.

**Black Notley** (655) is a parish two miles south from Braintree. Its church is a small flint building, and has a spire supported on massive timbers. This parish is notable as being the birthplace of John Ray, the naturalist, who was born here in 1628. His monument in the churchyard still stands. (p. 150.)

**Bocking** (3347) is a large parish on the Blackwater. Its church, situated on an eminence, is a fine building of flint and stone. Bocking has corn-mills and a famous crape factory. (pp. 66, 77, 78, 128.)

**Boreham** (820) is an interesting parish, with a picturesque church, which dates from the early thirteenth century. *New Hall*, in this parish, one of the grand Essex mansions, is a red brick building in the Tudor style. It was once a royal residence, where Henry VIII, Mary, and Elizabeth lived. (p. 126.)

**Bradwell-near-the-Sea** (783) is supposed to be on the site of the Roman station of *Othona*. The ancient church of St Peter-on-the-Wall probably dates from the early days of the Saxon invasion, and was the starting point of the work of Bishop Cedd. (pp. 86, 103, 107.)

**Braintree** (5330), a market town on the River Blackwater, was once a seat of the Bishops of London. The Flemings introduced the woollen trade in the reign of Elizabeth. This industry flourished till the end of the eighteenth century, when its place was taken by the weaving of silk and crape, which gives employment to hundreds of people in the town. (pp. 66, 75, 76, 77, 78, 99, 101.)

**Brentwood** (4932) is on the main Colchester and London road. It is of some antiquity and has a Grammar School, dating from 1557. The Essex Lunatic Asylum, a large building in beautiful grounds, is in this town. Brentwood is situated on a considerable eminence, and the surrounding country is well wooded and pleasant. (pp. 23, 32, 92, 94.)

**Brightlingsea** (4501), a fishing town and yachting centre, is situated on the east side of the Colne estuary. The church, a large and fine building, occupies an elevated situation and forms an important sea-mark. Some boat-building is carried on, and the people are largely engaged in the oyster fisheries. (pp. 24, 45.)

**Broomfield** (911), two miles north of Chelmsford, has a large village green round which the houses and cottages are picturesquely grouped. Its church, of great antiquity and interest, has a round tower. (p. 104.)

**Buckhurst Hill** (4786), pleasantly situated on a hill-top, has fine views over Epping Forest, and the Roding valley.

**Burnham** (2919) stands on the River Crouch. It has boat, barge, and ship building, and is one of the best yachting stations on the east coast. The oyster fisheries are extensive and of considerable value. (pp. 50, 73, 80, 83.)

**Burstead, Great** (1859) is the mother-parish of Billericay. The church is a large building of stone and rubble, the walls being of Norman age. Remains of a Roman encampment have been traced at a farm, one mile from Billericay.

**Castle Hedingham** (1097), on the River Colne, five miles west from Halstead, is a large and attractive parish. It is famous for its Norman Keep, perhaps the most perfect example in England. The church is a fine structure and has memorials of the De Veres, who for centuries held sway over this district. (pp. 75, 78, 92, 117, 119.)

**Chadwell St Mary** (5203) overlooks the marshes along the Thames. In this parish are the Tilbury Docks, belonging to the East and West India Dock Company. They are the finest in the kingdom and are capable of accommodating the largest vessels. (p. 40.)

**Chelmsford** (12,580), the county town of Essex, is situated in the valley of the Chelmer, near its junction with the Cann. It has considerable trade in agricultural produce, and there are extensive manufactures of agricultural implements and electrical apparatus. The chief buildings are the Church, Shire Hall and Corn Exchange, and the Grammar School founded by Edward VI. (pp. 3, 25, 62, 77, 79, 104, 136, 138.)

**Chesterford, Great** (785), in north-west Essex, was a Roman station of great importance, and many Roman remains have been found here. (p. 26.)

**Chigwell** (2508) is a pleasant parish on the borders of Epping Forest. Its Grammar School was founded in 1629, and has since been much enlarged. In the "King's Head Inn" is a large room in which Dickens wrote portions of *Barnaby Rudge*. (pp. 128, 144, 147, 151.)

**Chingford** (4373), a parish in the south-west of Essex, is much frequented by tourists on account of its proximity to the best parts of Epping Forest. The ruins of its old ivy-covered church are picturesque; and Queen Elizabeth's Hunting Lodge dates from Tudor times. (pp. 21, 140.)

**Clacton, Great** (7456) is a watering-place on the east coast, and has admirable facilities for bathing. (pp. 32, 46, 58, 59, 95, 115.)

**Coggeshall** (2882) is a very ancient town in the north-east of ·Essex, and has ruins of a priory. There are manufactures of silk, velvet, and isinglass. (pp. 77, 78, 104, 106, 108, 116, 128.)

**Colchester** (38,373) is a borough and market town on the Colne. It was a stronghold of the Britons; the *Colonia* of the Romans; and the *Colneceaster* of the Saxons. The Roman walls are almost entire, and Roman remains have been found in great numbers. The castle is the largest Norman Keep in England; and there are ruins of two abbeys. Colchester is the centre of an agricultural district, and its oyster fishery is extensive and famous. It has a good High Street with some fine public buildings; and a large Town Hall, opened in 1902. Colchester is the centre of a military district and has modern cavalry and artillery barracks. (pp. 66, 73, 77, 79, 83, 85, 86, 87, 92, 94, 98, 99, 101, 103, 108, 113, 117, 118, 136, 142.)

**Dagenham** (6091) is a very large parish, one mile east of Barking. In 1707, a high tide broke through the walls of the Thames, flooding 1000 acres of rich land and washing away 120 acres into the river. The breach was not stopped till 25 years later, and a large pond, near the site of the breach, is now preserved as a fishery. (pp. 39, 54, 78.)

**Danbury** (849) is about four miles south-east of Chelmsford. On the hill are remains of a camp, probably Danish. Danbury Park was once the residence of the Bishops of Rochester. (pp. 15, 95.)

**Dedham** (1500), a small quaint town on the Stour, was once a seat of the woollen trade. The church is large and fine, dating from the reign of Henry III. Dedham was the residence of Constable the landscape painter. An interesting group of houses, formerly the residence of the immigrant Flemings engaged in the "bay and say" (weaving) industries, still stands. (pp. 76, 77, 108, 110, 128, 151.)

**Dovercourt** (3894), a suburb of Harwich, is a pleasant watering-place. Its church has a lych-gate given by Queen Victoria, and a stained-glass window the gift of the German Emperor. (pp. 47, 56, 75.)

**Dunmow, Great** (2704) is a large parish and market town on the right bank of the Chelmer. (p. 22.)

**Dunmow, Little** (265), on the left bank of the Chelmer, was formerly the seat of a priory, which was remarkable for the ancient custom of the "Dunmow Flitch." (p. 116.)

**East Ham** (96,018) is a modern borough in the south-west of Essex. Its church is ancient and interesting; and the Town Hall and Technical Institute are fine modern buildings. (pp. 6, 68, 102, 108, 136.)

**Epping** (3789), at the north of Epping Forest, was long famous as a coaching town. It sends much dairy produce to London. Its High Street, nearly a mile long, has many good houses and shops, and the inns, with their old swinging signboards, date from the eighteenth century. (pp. 17, 32, 36, 73, 94, 98, 140, 146.)

**Felstead** (1945) is a large village, with some good houses, overlooking the Chelmer valley. The church is a large stone building with a Norman tower. The Grammar School, founded in 1554, holds a high position in the county.

**Frinton** (644) is a rising watering-place on the east coast. Its cliffs are 40 to 50 feet in height, and the beach is specially suitable for bathing. (p. 32.)

**Grays Thurrock** (13,834) is an important town on the Thames. There are extensive chalk quarries and cement works and in the Thames, off Grays, is moored the training-ship *Exmouth*. (pp. 30, 31, 33, 40, 77, 78, 100.)

**Greenstead** (110), near Ongar, is remarkable for its ancient church, mainly built of timber. (p. 109.)

**Hadleigh** (1343) is an ancient parish on the Thames. Its church of stone is Norman; and the castle, of which the ruins are on the summit of a hill, was built by Hubert de Burgh. (pp. 73, 117, 120.)

**Halstead** (6073) is a market town on the River Colne. The church of flint and stone dates from the fourteenth and fifteenth centuries; and the Grammar School was founded in 1594. There are manufactures of silk and crape. (pp. 66, 75, 76, 77, 78, 126.)

**Harlow** (2619) is an ancient market town on the River Stort. It was once of some importance, and is now noted for its annual fair.

**Harwich** (6176) is a borough, watering-place, and seaport in the north-east of Essex. It has been a place of note from early times, and its harbour is one of the best and safest on the east coast. There is regular daily communication between Harwich and the continent. The people are engaged in the shipping and fish trades, and there are manufactures of Roman cement and artificial manure. (pp. 20, 31, 47, 52, 56, 58, 75, 76, 80, 82, 136.)

**Hornchurch** (6402) is situated two miles south-east of Romford. Its church has a life-size representation in stone of a bullock's head, with two real horns inserted. Hornchurch has manufactures of brick, tiles, and drain-pipes.

**Ilford** (41,234) is a large and rapidly growing town on the River Roding. St Mary's Hospital is an interesting building founded in the reign of Henry II. Ilford is now famous for its extensive manufacture of photographic plates. (pp. 6, 22, 33, 78, 98.)

**Ingatestone** (1748) is an ancient little town, six miles south-west of Chelmsford. Its church has a fine embattled tower of red brick. Ingatestone Hall, in this parish, originally a quadrangular building, is in the Elizabethan style. (p. 110.)

**Kelvedon** (1569), a large village, stands on the Blackwater. Charles Spurgeon the nonconformist preacher was born here. (p. 144.)

**Leigh** (3667) is an ancient town, on a creek of the Thames, three miles west from Southend. Its church with a fine tower has a striking effect, and occupies a commanding position on the edge of the cliff. The town is a fishing station, and has considerable trade in oysters, shrimps, and cockles. (pp. 42, 80, 83.)

**Leyton** (98,912) is an ancient parish on the Lea. Its population is rapidly increasing, but there is little of interest in the place. The Town Hall and Technical Institute are good modern buildings. The Essex County Cricket Ground is at Leyton. Lodge the dramatist, Strype the historian, and Bowyer, the celebrated printer, were residents at Leyton. (pp. 6, 68, 98, 103, 144, 145.)

**Loughton** (4730) is a picturesque parish on the east of Epping Forest. In the neighbourhood is Loughton Camp, which was probably a British earthwork. (pp. 31, 99.)

**Maldon** (5565) is a borough, market town, and river port at the influx of the Chelmer to the Blackwater. It is of great antiquity, and suffered much at the hands of the Danes. There is some shipping trade. The oyster fisheries are of importance, and there are industries in boat-building, iron-founding, and brewing. (pp. 45, 79, 89, 110, 136.)

**Manningtree** (872), on the River Stour, is a market town with trade in malt, timber, and corn. There is also a large xylonite factory. (pp. 20, 47, 79.)

**Mistley** (1656) on the River Stour, a little east of Manningtree, has an extensive quay and considerable trade in corn, coals, etc.

**Ongar** (1117) is a small market town on the Roding. It was the site of a Roman settlement, and there are remains of a castle. (pp. 109, 117, 120.)

**Prittlewell** (27,245) is an ancient and pleasant parish one mile north of Southend, part of which is included in this place. The church is a very large and fine building, with a lofty tower of flint and stone chequer-work. (p. 110.)

**Rainham** (1725) is a large village on a creek of the Thames. It has a good quay, and some trade. Its church is a fine building dating from about 1155. (pp. 23, 78.)

**Rayleigh** (1773), an ancient town four miles north-east of Benfleet, has the remains of some earthworks which formed the site of a castle probably built before the Conquest. (p. 120.)

**Rochford** (1829) is a small market town of some antiquity, on the Roche, a branch of the Crouch. Rochford Hall, the remains of a large and fine mansion, was the residence of the Boleyn family, and the birthplace of Anne Boleyn, one of the wives of Henry VIII. (p. 128.)

**Romford** (13,656) is a market town of some importance. It has corn and cattle markets and is noted for its brewery. The Roman station *Durolitum* was probably here. (pp. 77, 94, 146.)

**Saffron Walden** (5896) is an ancient municipal borough and market town in the north-west of Essex. Many Roman remains have been found here, and there are earthworks, which were probably the site of a Roman encampment. The church, a large and magnificent stone building in the Perpendicular style, occupies a commanding position, and has a lofty crocketed spire. The Museum is one of the best of its kind, and contains an almost complete collection of British birds. *Audley End House*, the finest mansion in Essex, is close to this town. (pp. 30, 73, 77, 103, 104, 108, 110, 122, 136.)

**St Osyth** (1404), ten miles south-east of Colchester, has oyster beds. Near the village are the Abbey and the church, both of which are ancient and interesting. (p. 115.)

**Shenfield** (1962) is a large village near Brentwood. Its common, covered with furze, is very extensive.

**Southend** (28,857) is a seaport, watering-place, and borough on the estuary of the Thames. It is a popular resort of Londoners during the summer and autumn months. Its pier, the longest in England, the cliff gardens, and its extensive promenade are favourite resorts of visitors. The town is rapidly rising in importance, and carries on an active coasting trade. Steamers ply daily to Sheerness, Gravesend, and London. The Technical College is one of the best modern buildings. (pp. 32, 42, 58, 80, 136.)

**Springfield** (3274) is a large and pleasant village forming a suburb of Chelmsford. Here is the county gaol with accommodation for about 500 prisoners.

**Thaxted** (1659), a small town on the Chelmer, is famous for its fine church, which is sometimes called the "Cathedral of Essex." This splendid structure, dating from the fifteenth century, is of stone, and has a fine tower, with an octagonal spire, 181 feet high. Thaxted has some good houses, and many quaint timbered ones. The Guild Hall, or Moot Hall, a building of great antiquity, is mainly of timber, and stands in what was once the market-place. Samuel Purchas, author of works of travel and exploration at the beginning of the seventeenth century, was a native of Thaxted. (pp. 104, 108, 110, 148.)

**Upminster** (1477) is a pleasantly situated parish in the most picturesque part of Essex. The church is of considerable interest, and the tower with spire is 90 feet high. Upminster Hall is an ancient timbered mansion, with mullioned bay windows, wide gables, and clustered chimneys. There are also some good modern mansions in the Elizabethan style. (p. 23.)

**Waltham Abbey** (6549) is a market town on the Lea. Of the old Abbey, the nave of the church and the Lady chapel remain. Waltham Abbey has manufactures of gunpowder, cordite, small arms, etc. A large part of Epping Forest, once called Waltham Forest, is in this parish. (pp. 21, 71, 73, 75, 77, 89, 94, 116, 140, 144.)

**Walthamstow** (95,131) is a very large parish on the borders of Epping Forest. Its church dates from the twelfth century, and the Monoux almshouses are of the early sixteenth century. This parish was once a very favourite place of residence for wealthy London merchants and business men, but it is now the home of a working-class population The large reservoirs of the Metropolitan Water Board are in this parish. George Gascoigne, the Elizabethan poet, lived in Walthamstow; Benjamin Disraeli was educated in one of the old mansions; and William Morris, the poet, was born in a house which was recently demolished. (pp. 6, 33, 68, 75, 79, 98, 99, 128, 142, 145, 146.)

**Walton-le-Soken,** or **Walton-on-the-Naze** (2014) is a rapidly growing watering-place on the north-east coast. It has a good sandy beach, and the cliffs abound in fossils. (pp. 32, 46, 52, 58, 98.)

**Wanstead** (9179) is a pleasant forest parish, with some good houses and several large benevolent institutions. Wanstead Park, a fine woodland with a large lake, is now open to the public. The famous Wanstead House stood in its grounds. Penn, the quaker, and Tom Hood, the poet, lived at Wanstead. (pp. 6, 33, 146.)

**West Ham** (267,358), one of the most populous boroughs in England, is on the River Lea. It contains Stratford, Canning Town, and Plaistow, and each of these divisions has a large industrial population. The works of the Great Eastern Railway are in this borough and give employment to several thousand hands. There are also large docks at Canning Town; and shipbuilding, soap, candle, jute, chemical, and indiarubber works in various parts. Stratford market is an important centre for the sale of fruit and vegetables. Among the chief public buildings are the Town Hall, and an excellent Technical Institute, Library, and Museum, forming a really fine block in an attractive style of architecture. (pp. 6, 68, 75, 136.)

**Witham** (3454) stands near the influx of Pods Brook into the Blackwater. It is a market town of great antiquity, and once had some important woollen manufactures. (p. 72.)

**Wivenhoe** (2560) is a river port on the Colne, four miles south-east of Colchester. Ship and yacht building is the chief industry and there is also some trade in oysters. (pp. 24, 45, 80.)

**Woodford** (13,798) is a picturesque parish on the borders of Epping Forest, over which it has fine views. It consisted mainly of three villages, Woodford Green, Woodford Wells, and Woodford Bridge, and there were many good houses of wealthy London merchants. George Herbert, the poet, lived at Woodford, and Sydney Smith, the wit, and Coventry Patmore, the poet, were both born here. (pp. 6, 146, 147.)

**Writtle** (2718), two and a half miles west from Chelmsford, is the largest parish in Essex. Its village, with many curious old houses, is built round the village green. It is a place of great antiquity, and Roman remains have been found in the neighbourhood. (p. 77.)

ENGLAND & WALES　ESSEX

Fig. 1.  Diagram showing the area of Essex compared
with that of England and Wales

ENGLAND & WALES　ESSEX

Fig. 2.  Diagram showing the population of Essex
compared with that of England and Wales

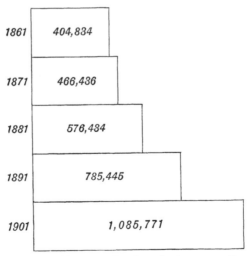

| 1861 | 404,834 |
| 1871 | 466,436 |
| 1881 | 576,484 |
| 1891 | 785,445 |
| 1901 | 1,085,771 |

Fig. 3.  Diagram showing the increase of
population from 1861—1901

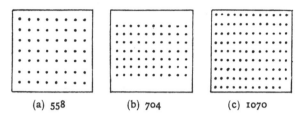

(a) 558          (b) 704          (c) 1070

Fig. 4. Diagram showing the density of population to a square mile in (a) England and Wales, (b) Essex, and (c) Lancashire. Each dot represents 10 people

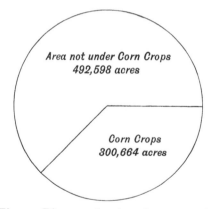

Fig. 5. Diagram showing the area under Corn Crops

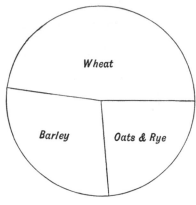

Fig. 6.  Diagram showing the proportionate areas
growing Wheat, Barley, Oats and Rye

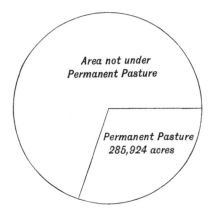

Fig. 7.  Diagram showing the area under
Permanent Pasture

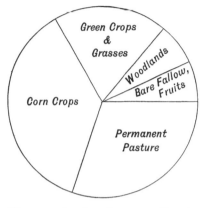

Fig. 8. Diagram showing the proportionate acreage
under Crops, Grass, etc.

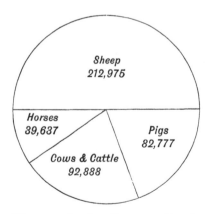

Fig. 9. Diagram showing the proportionate number
of sheep, cows and cattle, pigs, and horses

Milton Keynes UK
Ingram Content Group UK Ltd.
UKHW041520181024
449640UK00009B/87

9 781107 685543